LAB WORKB

FOUNDATIONS *of* ENGINEERING & TECHNOLOGY

R. Thomas Wright • Scott Bartholomew
Greg J. Strimel • Michael E. Grubbs

Seventh Edition

Publisher
The Goodheart-Willcox Company, Inc.
Tinley Park, IL
www.g-w.com

Manufactured in the United States of America.

ISBN 978-1-63126-888-5

1 2 3 4 5 6 7 8 9 – 19 – 22 21 20 19 18 17

Front cover images: RAGMA IMAGES/Shutterstock.com (left); LALS STOCK/Shutterstock.com (center); asharkyu/Shutterstock.com (right); Lena Serditova/Shutterstock.com (background)

Introduction

This lab workbook is designed for use with the ***Foundations of Engineering & Technology*** textbook. The chapters in the workbook correspond to those in the textbook and should be completed after reading the appropriate textbook chapter.

Each chapter of the workbook reviews the material found in the textbook chapters to enhance your understanding of textbook content. The various types of questions include matching, true or false, multiple choice, fill-in-the-blank, and short answer.

The lab workbook chapters also contain activities related to textbook content. The activities range from content reinforcement to real-world application, including design projects and broader modular activities. In these activities, it is important to understand any safety procedures set forth by your teacher.

Reading ***Foundations of Engineering & Technology*** and using this lab workbook will help you acquire a base of knowledge related to the principles of technology and engineering systems, as well as the design and application of each. Completing the questions and activities for each chapter will help you master the technical knowledge presented in the textbook.

Contents

Chapter Review Questions

Engineering and Technology Activities

Name ______________________________ Date __________ Class ________________

CHAPTER 1

Technology: A Dynamic, Human-Created System

Complete the following questions and problems after carefully reading the corresponding textbook chapter.

_______ 1. The application of knowledge, tools, and skills to solve problems and extend human capabilities is called ______.

A. development
B. information
C. literacy
D. technology

____________________ 2. *True or False?* People develop technology to modify or control the environment.

____________________ 3. The application of technology results in human-made things called ______.

____________________ 4. Technological innovation can also be called ______.

5. What does it mean when it is said that *technology is a dynamic process?*

____

____

____

6. Name three positive aspects of technology.

____

____

____

______ 7. Which of the following is *not* a potential negative aspect of technology?
A. Air pollution
B. Soil erosion
C. Tornadoes
D. Unemployment

______ 8. People who have an understanding of new technology and the ability to direct it are considered to be technologically ______.

______ 9. *True or False?* Early humans lived in civilized conditions.

10. Explain the relationship that the people who lived in primitive conditions had with nature.

______

______

______

______ 11. People are allowed to exert their will on the natural scene in ______ conditions.

Match the divisions in the history of civilization to the corresponding technological developments.

A. Bronze Age
B. Industrial Revolution
C. Information age
D. Iron Age
E. Middle Ages
F. Renaissance
G. Stone Age

______ 12. Copper tools, smelting, frescoes, writing paper, ink

______ 13. Desktop computers, robots, solar energy, cell research, satellites

______ 14. Fire, cave paintings, pottery

______ 15. Furnaces, aqueducts, body armor, ox-drawn plows, spinning wheels, windmills

______ 16. Printing press, magnetic compass, paper money, waterwheel

______ 17. Steam engine, cotton gin, power looms, factories, electricity, automobile, airplane

______ 18. Telescope, hydraulic press, calculating machine, modern architecture

______ 19. The ______ Age is the earliest period of known civilization.
A. Bronze
B. Fire
C. Iron
D. Stone

__________________ 20. Humans developed large-scale irrigation systems, writing, and navigation during the ______.

__________________ 21. *True or False?* Copper and bronze replaced iron and steel as the primary materials for tools because they are more plentiful and less costly.

______ 22. Which of the following developments of the Middle Ages was invented by Johann Gutenberg?
A. The magnetic compass
B. Paper money
C. The printing press
D. The waterwheel

__________________ 23. *True or False?* The Industrial Revolution is known as a period of new ideas in art, literature, history, and political science.

24. Name the period of civilization that is characterized by the introduction of and improvements to power-driven machinery, as well as new developments in manufacturing processes.

____

______ 25. Which of the following is *not* a characteristic of continuous manufacture?
A. The classification of management as a professional group
B. The combination of several tasks into the job of a single production worker
C. The creation of material-handling devices that bring the work to the workers
D. Improved machine life because of interchangeable parts

__________________ 26. Technology is now moving us into a new period, the ______, which is characterized by computer-based advances.

__________________ 27. *True or False?* In the information age, the previously sharp line between production workers and managers has begun to blur.

28. Name the type of technology used in growing food and producing natural fibers.

____

______ 29. ______ are used in processing data.
A. Agricultural and related biotechnologies
B. Communication and information technologies
C. Energy and power technologies
D. Manufacturing technologies

__________________ 30. ______ technologies are used in building structures for housing, business, transportation, and energy transmission.

__________________ 31. *True or False?* Transportation technologies are used in converting and applying energy to power devices and systems.

32. Name the type of technology used in converting materials into industrial and consumer products.

____

_______ 33. ______ are used in maintaining health and curing illnesses.
A. Agricultural and related biotechnologies
B. Communication and information technologies
C. Energy and power technologies
D. Medical technologies

_______________ 34. ______ technologies are used in moving people and cargo from one place to another.

_______________ 35. *True or False?* On its way from its natural state to a finished product, a material might be involved with several different types of technology.

36. Name the type of technology involved with a product's sales orders, engineering drawings, specifications sheets, advertising, and sales reports.

____

_______ 37. ______ help to power furnaces and presses, light work areas, and heat offices and control rooms.
A. Agricultural and related biotechnologies
B. Communication and information technologies
C. Energy and power technologies
D. Manufacturing technologies

_______________ 38. ______ and related technologies facilitate the growth of trees.

_______________ 39. *True or False?* Medical technologies help to ensure that workers remain in good health.

40. Name the type of technology essential for creating access to mines and quarries, travel on roads, power lines, and dams and pipelines.

____

_______ 41. ______ move raw materials and industrial materials and deliver finished products.
A. Agricultural and related biotechnologies
B. Energy and power technologies
C. Manufacturing technologies
D. Transportation technologies

Name ______________________ Date __________ Class ______________

CHAPTER 2

Connecting Technology and Engineering through Mathematics and Science

Complete the following questions and problems after carefully reading the corresponding textbook chapter.

________________ 1. ______ is the development and application of knowledge, tools, and human skills to solve problems and extend human potential, brought about by needs and desires.

________________ 2. *True or False?* Humans are the only animals that can use tools.

________________ 3. ______ is the measure of the ability to amplify the amount of effort exerted using some type of mechanical device.

______ 4. The claw of a hammer is used as which type of simple machine?
A. Inclined plane
B. Lever
C. Pulley
D. Wedge

______ 5. Who are the people who conduct research and apply scientific and technological knowledge to the design and development of products, structures, and systems to solve real-world problems?
A. Engineers
B. Mathematicians
C. Scientists
D. Technologists

Match the fields of study to the corresponding letters in the following figure.

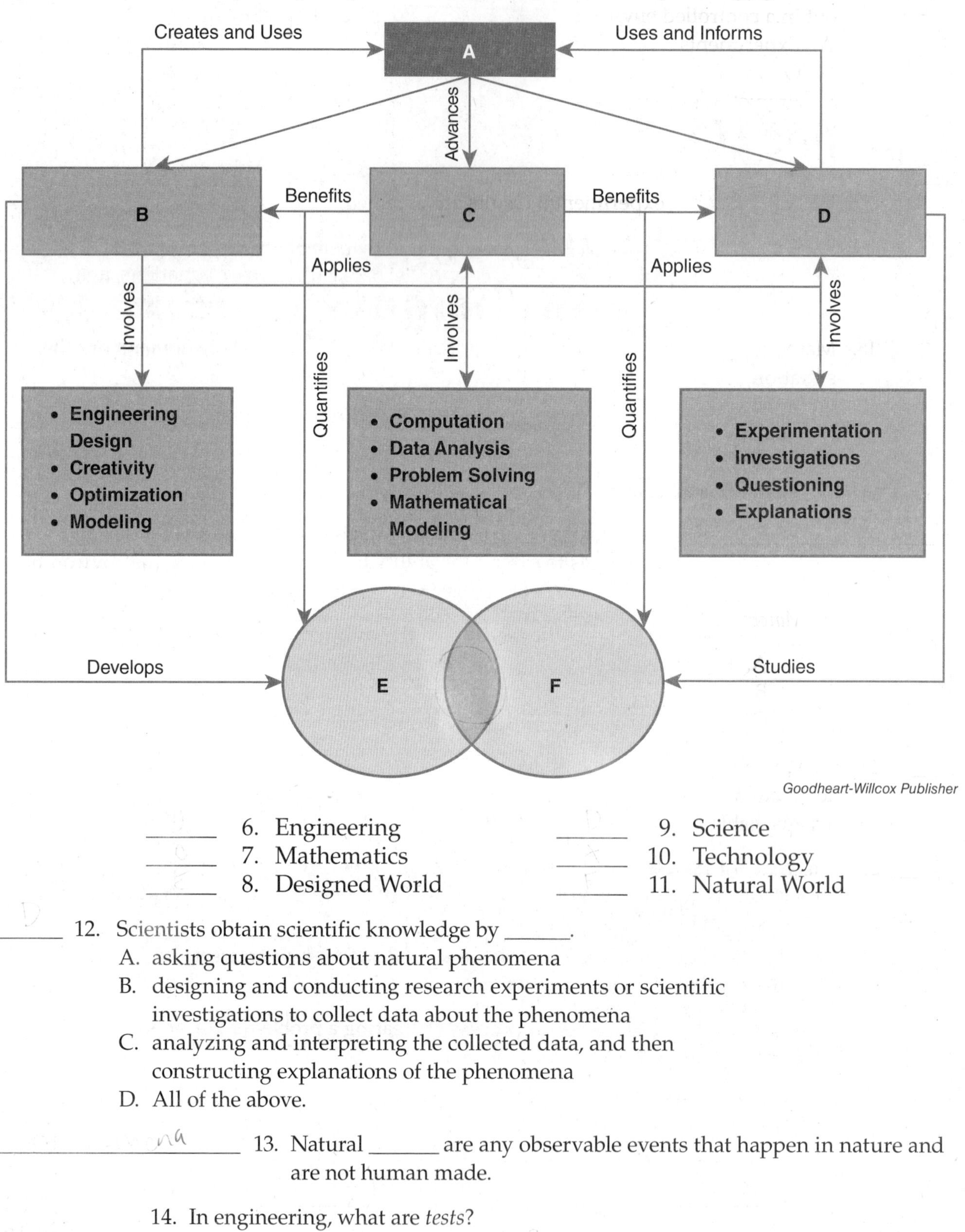

Goodheart-Willcox Publisher

_____ 6. Engineering
_____ 7. Mathematics
_____ 8. Designed World
_____ 9. Science
_____ 10. Technology
_____ 11. Natural World

_____ 12. Scientists obtain scientific knowledge by _____.
A. asking questions about natural phenomena
B. designing and conducting research experiments or scientific investigations to collect data about the phenomena
C. analyzing and interpreting the collected data, and then constructing explanations of the phenomena
D. All of the above.

_____ 13. Natural _____ are any observable events that happen in nature and are not human made.

14. In engineering, what are *tests*?

____

____

____

Name ____________________________________

______ 15. Which of the following are established research procedures carried out in a controlled environment to prove or disprove a hypothesis?
A. Experiments
B. Optimizations
C. Predictive analyses
D. Tests

______________________ 16. The ______ method structures research in a way that ensures valid experimental results are obtained.

______________________ 17. *True or False?* Science is a tool that enables people to quantify the world by expressing situations as numbers, equations, and inequalities.

______ 18. Mathematics allows scientists and engineers to describe natural phenomena or other situations as ______.
A. hypotheses
B. qualities
C. quantities
D. technological designs

______________________ 19. Technology ______ and uses tools and machines, systems, and materials to extend the ability to control and modify the environment.

Match the term to its description.

A. Engineering
B. Hypothesis
C. Mathematics
D. Science
E. Technological design

______ 20. The practical application of science, mathematics, and technological know-how to solve problems in the most efficient way possible

______ 21. The study of the natural world

______ 22. A proposed explanation of a situation

______ 23. The study of measurements; patterns; and the relationships between quantities, using numbers and symbols

______ 24. The open-ended, trial-and-error process of creating a problem solution to meet needs and desires

25. What is the purpose of predictive analysis?

____

____

____

______ 26. ______ is the engineering practice of making something as fully effective or as perfect as possible using mathematical procedures.
A. Innovation
B. Mechanical advantage
C. Optimization
D. Predictive analysis

______ 27. *True or False?* The first step of the design process is to gather information.

______ 28. In the design process, which of the following is *not* part of developing a design solution?
A. Detailing the best solution
B. Isolating the best solution
C. Refining several possible solutions
D. Searching for all necessary background information

29. In the design process, what are three types of models or prototypes that can be produced?

______

______

______ 30. In the design process, people ______ the solution by testing a model of the selected solution to determine how well it actually solves the problem.

______ 31. *True or False?* When communicating the final solution during the design process, a final solution is selected, and documents needed to produce and use the device or system are prepared.

32. What is *technology transfer*?

______

______

______

______ 33. A new, useful product or process that solves some type of problem and that did not previously exist is a(n) ______.

______ 34. ______ are refinements or improvements made to preexisting products or processes that better solve a problem.

35. What is a *patent*?

______

______

______

Name ______________________ Date ____________ Class ______________________

CHAPTER 3

Engineering Fundamentals

Complete the following questions and problems after carefully reading the corresponding textbook chapter.

______ 1. Engineering as a profession can be traced back to as early as ______.
   A. 6000 BCE
   B. 30 BCE
   C. the Renaissance
   D. the twentieth century

______________ 2. *True or False?* The current engineering profession takes a more organized and disciplined approach to solving problems than the trial-and-error approach of the past.

______________ 3. ______ engineering focuses on the design and improvement of structures such as bridges, skyscrapers, roadways, and dams.

4. What is *electrical engineering*?

____

____

____

______ 5. Aerospace engineering and biomedical engineering are considered to be ______ branches of engineering.
   A. chemical
   B. environmental
   C. interdisciplinary
   D. mathematical

______________ 6. *True or False?* Engineers tend to have extensive background knowledge in science, mathematics, and technical disciplines and be problem solvers and logical thinkers.

7. What does the term *iterative* mean?

____

____

_____ 8. Which of the following is the best meaning for the term *optimize*?
 A. Constrain
 B. Improve
 C. Limit
 D. Repeat

_______________ 9. *True or False?* Some common engineering characteristics include thinking like everyone else, working independently, and pessimism.

_______________ 10. _______ is important because most engineering design projects are undertaken as a team.

11. Name three means through which ideas and outcomes can be communicated.

____

____

____

_______________ 12. The principles of conduct that govern the actions of an individual or a group are _______.

_______________ 13. *True or False?* The Engineers' Creed contains a preamble, fundamental canons, rules of practice, and professional obligations.

14. What is a *creed*?

____

____

____

_____ 15. A set of interacting elements that work interdependently to form a complete entity that has an overall function or purpose is a(n) _______.
 A. creed
 B. dimension
 C. system
 D. unit

_______________ 16. *True or False?* Making estimations, determining significant figures, and using scientific and engineering notation are some of the basic skills that effective engineers use throughout their careers.

_______________ 17. Engineers must be able to make accurate _______, or rough calculations, because, often, insufficient information is provided to design a solution.

Name Jessi ______________________

18. What are *significant figures?*

____

____

____

Match the terms to their corresponding descriptions.

A. Chemical engineering
B. Creativity
C. Engineering design
D. Mechanical engineering
E. Optimism
F. Scientific notation

_______ 19. An engineering discipline that involves working with the assembly of parts

_______ 20. An engineering discipline that involves the conversion of raw materials into useful products

_______ 21. A problem-solving process that involves creating solutions to problems following specified criteria (guidelines) and under specific constraints in an iterative manner

_______ 22. Thinking in a way that is different from the norm

_______ 23. The tendency to look at the more favorable side of events or expect the best outcomes in various situations

_______ 24. A way to express numbers that includes just the digits of a number with the decimal point placed after the first digits, followed by a multiplication of a power of ten that will put the decimal back in the correct place

_______ 25. Which of the following is a modification of scientific notation in which all the powers of 10 are multiples of 3?
A. Engineering notation
B. Estimation
C. Mathematical notation
D. Technological notation

______________ 26. A(n) _______ chart is a tool used to illustrate a project schedule using horizontal bars representing each phase and activity making up the project.

_______ 27. Which of the following is a form of expository, informative, or explanatory writing used to communicate complex information to those who need it for a specific reason?
A. Engineering notation
B. Scientific writing
C. Technical notation
D. Technical writing

______________ 28. *True or False?* Presentations must be balanced with audio elements.

______________ 29. Newton's laws of motion and the laws of thermodynamics are examples of _______ laws that engineers must understand.

A 30. What type of graphics and drawings are almost always created as or transferred to digital graphics or models using three-dimensional design software?
(A) Engineering
B. Iterative
C. Mathematical
D. Scientific

31. What is a *design dimension*?

A measurable feature of an object such as length, width, or height.

unit 32. A physical quantity that is consistent to a standard form of measurement is called a(n) ______.

True 33. *True or False?* Engineers use data analysis software to record, organize, and analyze data when conducting experiments and tests.

Name ______________________________ Date __________ Class ____________________

CHAPTER 4

Engineering Design Problem Solving

Complete the following questions and problems after carefully reading the corresponding textbook chapter.

______ 1. ______ is a thinking and reasoning process that can be applied to a broad range of situations.
A. Marketing
B. Problem solving
C. Production
D. Sustainability

____________________ 2. *True or False?* Design is the same as problem solving.

____________________ 3. The ______ process requires people to employ thinking and reasoning skills to consider and find solutions for a situation that has an identified goal state, readily available solutions, and one or more reasonable pathways for solving.

4. What are the four main steps used in problem solving?

____

____

____

____

Match each problem variation to its description.

A. Complexity
B. Context
C. Structure

______ 5. How complicated and difficult a problem is to solve

______ 6. How much information is provided, implying how a successful solution will look

______ 7. The set of circumstances or facts surrounding a specific problem

_______ 8. What kind of design involves the arrangement of components to produce a desired result?
A. Concept design
B. Product design
C. Sustainable design
D. System design

_______________ 9. In construction, _______ design involves the development of a structure to meet the needs of customers.

_______________ 10. *True or False?* A product or structure must be designed for function, production, and marketing.

11. What is meant when it is said that a product must be *designed for function*?

____

____

____

_______________ 12. A structure must be designed for _______, meaning it must be easy to construct.

_______________ 13. *True or False?* A product or structure must be appealing to the designer.

_______________ 14. _______ in engineering involves product and system design, development, integration, and implementation.

15. Briefly describe *engineering design.*

____

____

____

16. List four types of restrictions that can be included in engineering design problem constraints.

____

____

____

____

_______ 17. _______ is a method used to solve problems by designing a product or system that meets a desired goal, while adhering to established constraints and taking into consideration many factors, such as potential impacts, risks, and benefits.
A. The engineering design process
B. Predictive analysis
C. The problem-solving process
D. Production strategy

________________ 18. *True or False?* The steps of the engineering design process always proceed in a direct path from start to finish.

_______ 19. Which of the following is *not* a task included in the *Identify and Define a Problem* step of the engineering design process?
A. Formulating a problem statement
B. Identifying and validating a problem
C. Researching current solutions
D. Understanding and establishing criteria and constraints

________________ 20. *True or False?* Separating the engineering design project into smaller, more manageable tasks and evaluating the resources available to develop a solution are beneficial strategies for defining the problem in the engineering design process.

21. Identify two tasks that should be completed during the *Gather Information* step of the engineering design process.

____

____

_______ 22. Which of the following is *not* a beneficial strategy for gathering information in the engineering design process?
A. Categorize the superficial details of the problem.
B. Draw inferences to other potential ideas or concepts using past and current solutions for similar problems.
C. Monitor solution progress based on the established benchmarks for project completion.
D. Recognize relevant information and organize it into patterns.

________________ 23. Searching for solutions; isolating, refining, and detailing the best solution; and finalizing design specifications are tasks included in the _______ *a Design Solution* step of the engineering design process.

________________ 24. *True or False?* A concept map is a tool you can use to graphically organize information.

________________ 25. A(n) _______ is a diagram that helps people visually organize information or ideas around a central topic or theme.

_______ 26. Which of the following is a chart used to record a rating for how well a design meets a desired criterion?
A. A bill of materials
B. A decision matrix
C. A Gantt chart
D. A mind map

________________ 27. *True or False?* Acquiring the appropriate resources, prototyping the solution, and modifying the design solution are tasks included in the Evaluate the Solution step of the engineering design process.

28. Describe three strategies that are beneficial for making a solution in the engineering design process.

____

____

____

____

____

____

______ 29. Which of the following is *not* a task included in the *Evaluate the Solution* step of the engineering design process?
- A. Conducting material experimentation
- B. Conducting mathematical examinations
- C. Determining test criteria
- D. Establishing a testing procedure

______________ 30. Developing a detailed description of the testing procedure, using software to help analyze testing data, and conducting a critical design review with external stakeholders are beneficial strategies for the ______ *the Solution* step of the engineering design process.

Match each step of the engineering design process to its description.

A. Communicate the final solution.
B. Develop a design solution.
C. Evaluate the solution.
D. Gather information.
E. Identify and define a problem.
F. Model and make a solution.

______ 31. Acquiring the basic information about the problem and establishing the design limitations

______ 32. Obtaining the knowledge necessary to develop a design solution by conducting research

______ 33. Creating and refining several possible solutions and detailing the best solution

______ 34. Producing physical, graphic, computer, or mathematical models of the selected solution

______ 35. Testing the physical, graphic, computer, or mathematical models of a selected solution

______ 36. Selecting a final solution and preparing the documents and presentations needed to share the evaluation outcomes and produce and use the final solution

______________ 37. *True or False?* The *Communicate the Final Solution* step of the engineering design process includes tasks, such as studying the solution results, sharing the results and conclusions, and presenting the solution for approval.

38. Describe two strategies that are beneficial for communicating results in the engineering design process.

____

____

____

____

__________________ 39. *True or False?* Many design actions are iterative and used throughout all steps of the design process.

Name ______________________________ Date __________ Class ____________________

CHAPTER 5

Developing Design Solutions

Complete the following questions and problems after carefully reading the corresponding textbook chapter.

Match the steps in the engineering design process to the corresponding letters in the figure.

Engineering Design Process

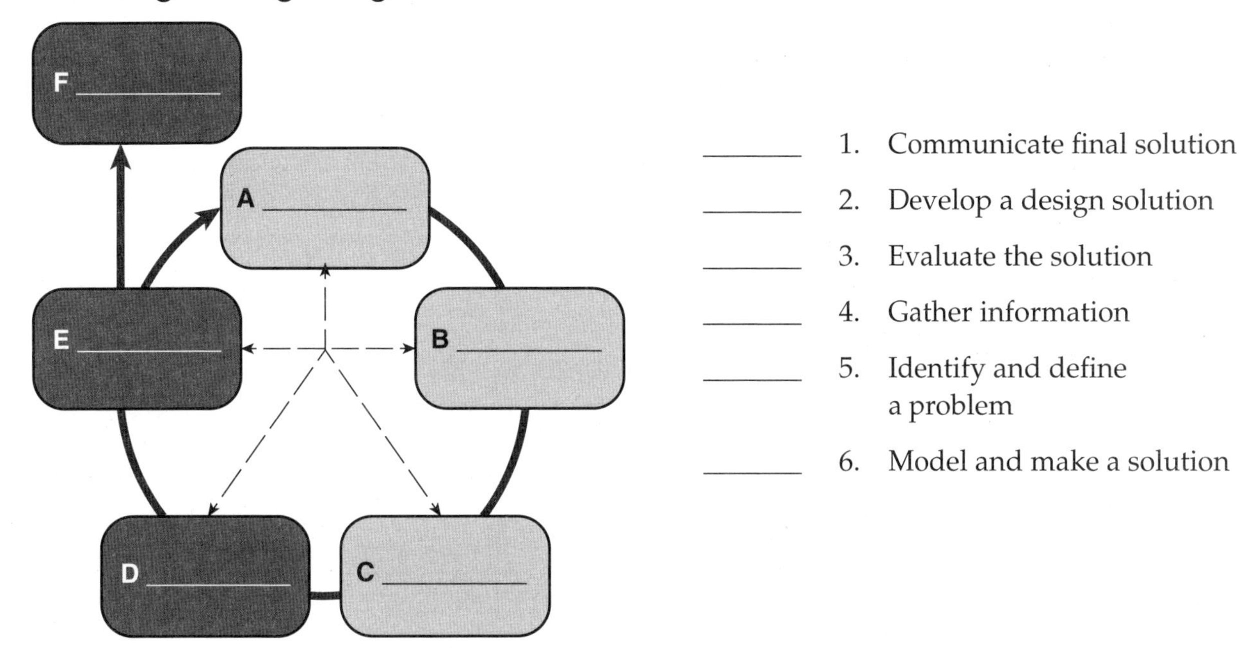

Goodheart-Willcox Publisher

_______ 1. Communicate final solution

_______ 2. Develop a design solution

_______ 3. Evaluate the solution

_______ 4. Gather information

_______ 5. Identify and define a problem

_______ 6. Model and make a solution

_______ 7. Using a _______ allows various people to contribute their talents to a project.

A. depth-first approach
B. design team
C. predictive analysis
D. synergism

_______________ 8. *True or False?* Identifying the problem is always the simplest step of the engineering design process.

_______________ 9. When describing a situation using the depth-first approach, engineers must consider multiple questions and might begin with a(n) ______ map to consider options.

______ 10. Which of the following categories of criteria and constraints describes the operational and safety characteristics the device or system must meet?

A. Financial
B. Market
C. Production
D. Technical, or engineering

_______________ 11. "Must be manufactured using existing equipment in the factory" is an example of a(n) ______ constraint.

_______________ 12. *True or False?* To solve a problem, historical information about devices and systems that have been developed to solve similar problems might be needed.

13. Name three types of information used as a foundation for technological-development activities.

____

____

____

______ 14. What do we call a decision that arises in which a choice must be made between two competing items?

A. A classification
B. A constraint
C. Synergism
D. A trade-off

_______________ 15. *True or False?* Brainstorming, classification, and what-if scenarios are three popular techniques for generating ideas.

16. What is *synergism*?

____

____

____

______ 17. Which of the following is *not* a good rule for brainstorming activities?

A. Encourage unconventional, unusual ideas.
B. Maintain a rapidly paced session.
C. Record the ideas without reacting to them.
D. Seek quality, not quantity.

_______________ 18. Incomplete, unrefined sketches used to communicate design solutions are called ______ sketches.

19. Describe the purpose of a detailed sketch.

____

____

____

____

20. What are two kinds of pictorial sketches designers produce when refining ideas?

____

____

_______ 21. A(n) _______ sketch shows the front view of an object as if a person is looking directly at it.
A. isometric
B. oblique
C. perspective
D. rough

_______ 22. In which of the following types of sketches are the angles formed by the lines in the upper-right corner of the object all equal?
A. Cavalier
B. Isometric
C. Oblique
D. Perspective

23. Briefly describe the steps used to create an isometric sketch.

____

____

____

____

____

_______________________ 24. A(n) _______ sketch shows the object as the human eye or a camera sees it, and in this type of sketch, lines are drawn to meet at distant vanishing points on the horizon.

_______________________ 25. *True or False?* Whether they are developing the basic structure for a one-, two-, or three-point perspective sketch, designers follow the same basic steps.

_______________________ 26. _______-view drawings are used to show the layout of flat pieces, which usually have standard or predetermined thicknesses.

_______________________ 27. *True or False?* In two-view drawings, the two views shown are typically the front and top views.

28. How many views are typically used to show the size and shape of rectangular and complex parts?

____

_______ 29. A multiview drawing that shows the top, right-side, and end views is called a third-angle _______.
   A. assembly
   B. drawing
   C. projection
   D. view

_______________________ 30. Additional sketched views of an object needed for complex parts are called _______ views.

_______ 31. Which of the following steps is *not* part of the correct technique for including dimensions on a drawing?
   A. Complete the side view by constructing the overall shape from the front and top projections.
   B. Dimension the locations of all features.
   C. Dimension the size of the object first, followed by the sizes of all major features.
   D. Indicate any necessary geometric dimensions.

Match the types of lines to the corresponding letters in the figure.

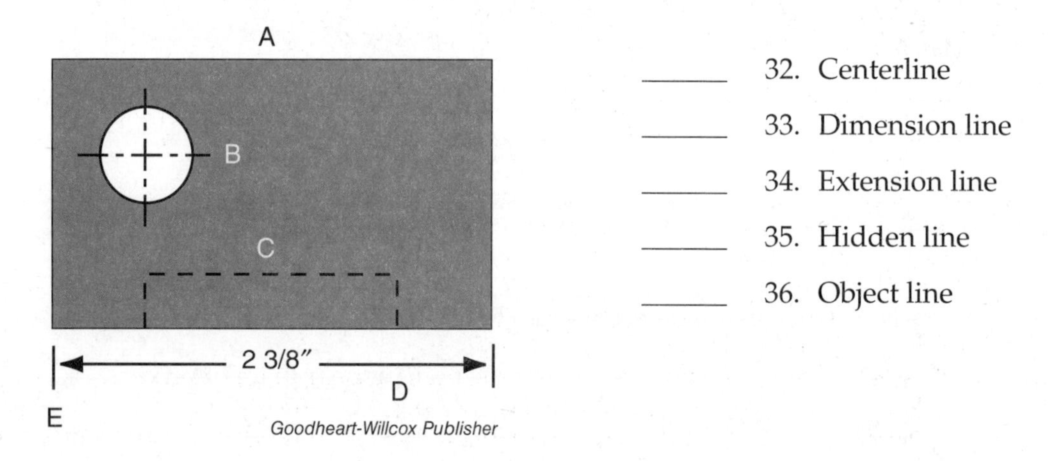

Goodheart-Willcox Publisher

_______ 32. Centerline

_______ 33. Dimension line

_______ 34. Extension line

_______ 35. Hidden line

_______ 36. Object line

_______ 37. Which type of drawing shows how parts fit together?
   A. Assembly drawing
   B. Orthographic drawing
   C. Pictorial drawing
   D. Systems drawing

_______________________ 38. *True or False?* Orthographic assembly drawings use standard orthographic views to show parts in their assembled positions.

39. What are *systems drawings*?

____

____

____

__________ 40. A simulation of actual events, structures, or conditions is called ______.

41. What is a *chart*?

____

____

____

__________ 42. *True or False?* Diagrams allow designers to organize and plot data.

______ 43. Which of the following types of models can test the strengths of materials and structures and can be used to observe a product during normal operation?
A. Computer models
B. Graphic models
C. Mathematical models
D. Physical models

Match each term to its description.

A. Diagram
B. Mathematical model
C. Scientific model
D. Solid model
E. Surface model
F. Wire-frame model

______ 44. A model that shows the relationships among components in a system

______ 45. A type of computer model developed by connecting all the edges of the object, producing a structure made up of straight and curved lines

______ 46. A computer model that shows how a project will appear to an observer

______ 47. A complex computer model that takes into account both the surface and interior substance of an object

______ 48. A model that shows relationships through formulas

______ 49. A simulation based on existing or accepted findings related to the natural world

______ 50. Which of the following tasks can occur at other stages of designing, but is especially important between the steps of developing and making design solutions?
A. Classification
B. Optimization
C. Predictive analysis
D. Production

Name ______________________ Date __________ Class ______________

CHAPTER 6

Making Design Solutions

Complete the following questions and problems after carefully reading the corresponding textbook chapter.

______________ 1. *True or False?* Designs are modeled in the *Develop a Design Solution* step of the engineering design process.

2. List two actions that are involved in the conversion of a technical drawing or 3-D computer model into a physical model.

____

____

____

____

______________ 3. ______ is the simulation of actual events, structures, or conditions.

______ 4. Which of the following is *not* a traditional type of physical model?

A. Mock-ups
B. Prototypes
C. Scale models
D. Squares

______________ 5. A(n) ______ is a three-dimensional miniature physical representation of a design solution that maintains accurate relationships between all aspects and features of the design.

6. What is a *prototype*?

____

____

____

______ 7. A focus on ______ is crucial when making physical models.

______ 8. Changing the form of materials is called materials ______.

9. Name two common characteristics of all machine tools.

______

______

______

______

Match the parts of the cutting tool to the corresponding letters in the figure below.

Goodheart-Willcox Publisher

______ 10. Rake angle

______ 11. Relief angle

______ 12. Sharpened edge

______ 13. A spindle, or shaft, used to hold table-saw blades and milling cutters is a(n) ______.

______ 14. Which of the following is a type of machine tool?
A. Sawing machine
B. Drilling machine
C. Planing machine
D. All of the above.

______ 15. A(n) ______ is a turning machine that produces cutting motion by rotating the workpiece.
A. arbor
B. broach
C. chuck
D. lathe

______ 16. *True or False?* The headstock in a metal lathe is called an arbor.

______ 17. Cutting along the length of a workpiece to produce a cylinder of uniform diameter is called ______.
A. facing
B. grooving
C. tapering
D. turning

______________ 18. Cutting along the length of a cylinder at a slight angle to produce a cylindrical shape with a uniformly decreasing diameter is called ______.

______________ 19. *True or False?* Cutting along the end of a rotating workpiece to produce a true end is called chamfering.

20. Briefly describe the lathe operation of grooving.

____

____

____

____

______________ 21. *True or False?* Cutting threads along the outside diameter of or inside a hole in a workpiece is called threading.

22. Briefly describe the lathe operation of knurling.

____

____

____

____

______ 23. ______ machines use teeth on a blade to cut material to a desired size and shape.

A. Drilling
B. Planing
C. Sawing
D. Shaping

Match the sawing operations to the corresponding letters in the figure below.

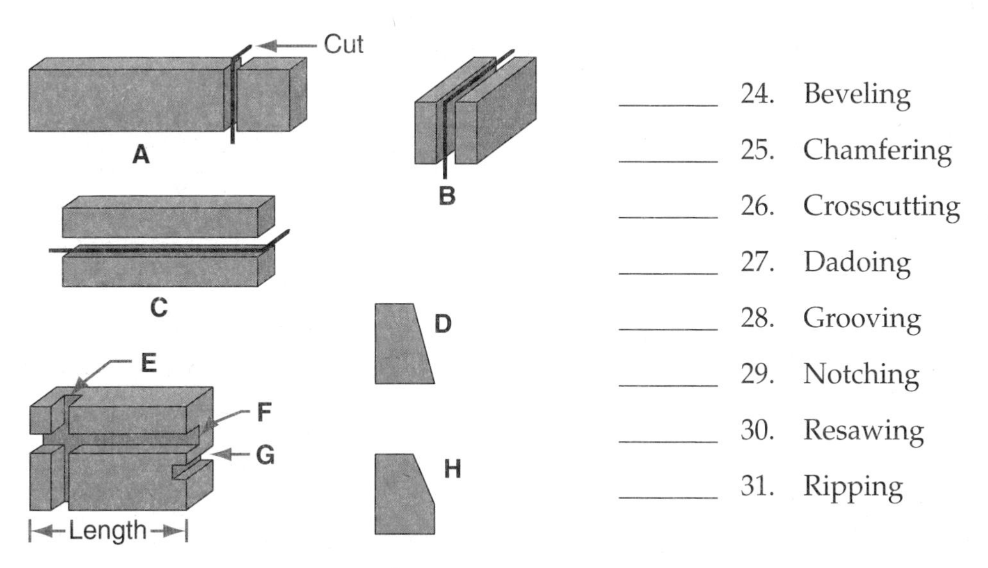

Goodheart-Willcox Publisher

______ 24. Beveling

______ 25. Chamfering

______ 26. Crosscutting

______ 27. Dadoing

______ 28. Grooving

______ 29. Notching

______ 30. Resawing

______ 31. Ripping

______________________ 32. ______ saws use a blade in the shape of a disk with teeth arranged around the edge.

______________________ 33. *True or False?* Scroll saws use a blade made of a continuous band of metal, usually with teeth on one edge.

34. What is a *drilling machine*?

____

____

____

______ 35. Which of the following is *not* a common drilling machine operation?
A. Counterboring
B. Countersinking
C. Reaming
D. Twisting

______________________ 36. *True or False?* Counterboring is a drilling machine operation that produces a beveled outer portion of a hole.

37. Briefly describe the drilling machine operation of reaming.

____

____

____

____

____

______ 38. A(n) ______ is a machine that uses a tool with many teeth, each tooth sticking out slightly more than the previous tooth.
A. broach
B. chuck
C. grinding machine
D. shaping machine

______________________ 39. A machine that uses a bonded abrasive to cut material is a(n) ______ machine.

40. What is a *surface grinder*?

____

____

____

____

____

____

_______ 41. ______ is the practice of comparing the qualities of an object to a standard.
A. Indirect reading
B. Material processing
C. Measurement
D. Prototyping

______________________ 42. The International System of Units (SI or *Système international dèunités*) is more commonly known as the ______ system.

______________________ 43. *True or False?* The US customary system is a measurement system used in the United States that includes units such as the inch, foot, mile, pound, pint, quart, and gallon.

_______ 44. In the metric system, 8000 meters are equal to 8 ______.
A. centimeters
B. hectometers
C. kilometers
D. millimeters

______________________ 45. *True or False?* In the metric system, 23 millimeters are equal to 23/100 of a meter.

46. Briefly explain standard measurement.

____

____

____

____

____

____

_______ 47. Measurement made to 1/1000″ to 1/10,000″ in the US customary system or to 0.01 mm in the metric system is called ______ measurement.
A. direct-reading
B. indirect-reading
C. precision
D. standard

______________________ 48. Used for linear measurement, a strip of metal, wood, or plastic with measuring marks on its face is called a(n) ______.

Match each tool to its description below.

A. Direct-reading measurement tool
B. Indirect-reading measurement tool
C. Micrometer
D. Square

_______ 49. A measurement tool that an operator manipulates and then reads

_______ 50. A measurement tool that automates measurement using sensors and computers

_______ 51. A precision measuring tool used to measure diameters

_______ 52. A tool with a blade that is at a right angle to the head

53. What is *quality control*?

____

____

____

_______ 54. Which of the following should be done during the *Model and Make a Solution* step of the engineering design process?
A. Communicate the solution
B. Gather information
C. Identify the problem
D. Optimize the solution

Name ______________________________ Date ___________ Class ____________________

CHAPTER 7

Evaluating Design Solutions

Complete the following questions and problems after carefully reading the corresponding textbook chapter.

______ 1. Tests, experiments, and analysis are part of the ______ *the Solution* step of the engineering design process.
A. *Communicate*
B. *Develop*
C. *Evaluate*
D. *Model*

2. Name four of the steps involved in the evaluation phase.

____

____

____

____

____________________ 3. Engineers identify test ______ through specification analysis, functional analysis, and design analysis.

____________________ 4. *True or False?* Tolerance is the amount a dimension can vary and still be acceptable.

Match each type of analysis to its corresponding description.

A. Design analysis
B. Ergonomics
C. Functional analysis

______ 5. A type of analysis that evaluates the degree to which a product, structure, or system operates effectively under the conditions for which it was designed

______ 6. Analysis that helps engineers, designers, and decision makers assess the design's purpose and likely acceptance in the marketplace

______ 7. The science that considers the size and movement of the human body, mental attitudes and abilities, and the senses

______ 8. Which of the following is *not* a type of design analysis?
A. Economic analysis
B. Functional analysis
C. Human-factors analysis
D. Market analysis

______________________ 9. ______ analysis includes determining customer expectations for the product's appearance, function, and cost and studying present and anticipated competition.

10. What is a *return on investment (ROI)*?

____

____

____

____

______________________ 11. *True or False?* An engineer might be required to follow technical standards, which are unique, custom plans.

12. What is a *physical test*?

____

____

____

____

______ 13. Which of the following is *not* a physical demand that is typically tested with a physical test?
A. Compression
B. Extended period of use
C. Tensile stress
D. Torsion

______________________ 14. A pulling motion in opposite directions of an object is called ______ stress.

______________________ 15. *True or False?* Compression is a twisting motion on an object.

16. What is a *performance test*?

____

____

____

____

_______ 17. Which of the following is *not* a step of the scientific method?
- A. Construct a hypothesis.
- B. Draw a conclusion.
- C. Model and make the solution.
- D. Test the hypothesis with an experiment.

______________________ 18. ______ methods of data collection involve numbers and objective data.

______________________ 19. *True or False?* Qualitative methods of data collection involve words or subjective data.

Match the type of graph to the corresponding letters in the figure below.

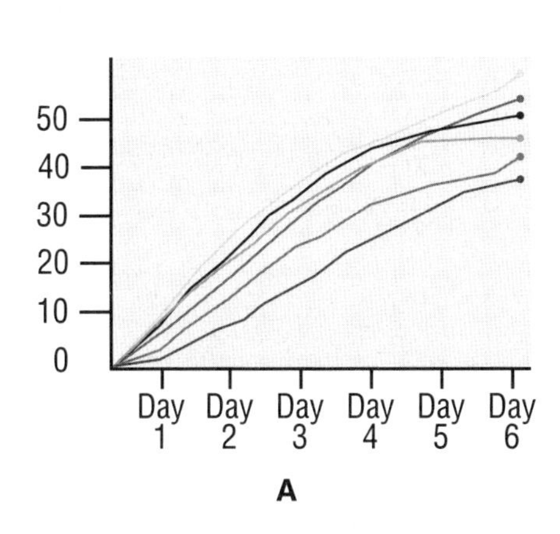

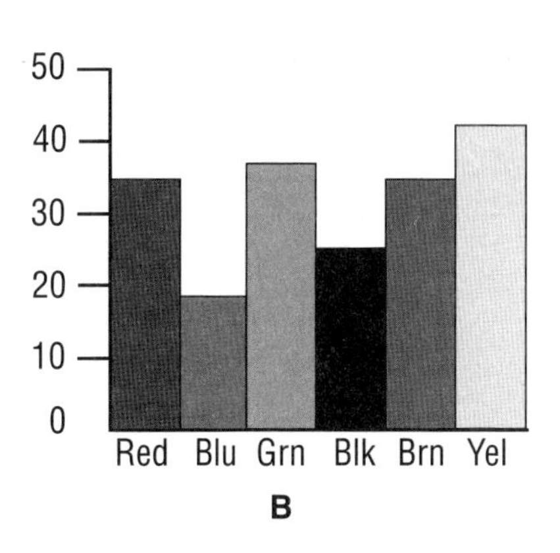

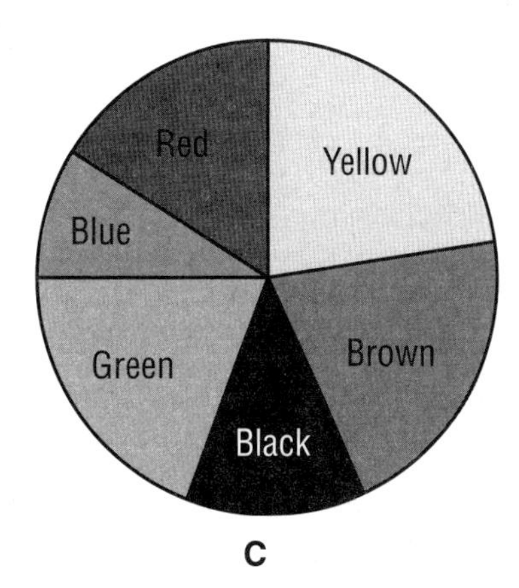

Goodheart-Willcox Publisher

_______ 20. Bar graph

_______ 21. Line graph

_______ 22. Pie graph

Match each type of graph or chart to its corresponding description.

A. Comparison/relationships
B. Composition charts
C. Distribution charts
D. Trend charts

_______ 23. Used when data is available for two or more variables. Examples include line charts, bar charts, and Venn diagrams.

_______ 24. Used to visualize how a group of data is dispersed throughout the group. Examples include column and line histograms and scatter plot charts.

_______ 25. Used to show how data changes over time. Examples include line charts and graphs.

_______ 26. Used to show how data is divided. Examples include pie charts.

_______ 27. When flaws of a design are identified, often, _______ is required.
A. nothing
B. redesign
C. starting from scratch
D. throwing the design out completely

Name ______________________________ Date __________ Class __________________

CHAPTER 8

Communicating Design Solutions

Complete the following questions and problems after carefully reading the corresponding textbook chapter.

______________________ 1. *True or False?* Communication takes place only at the end of the engineering design process.

2. What is a *stakeholder*?

____

____

____

______ 3. A(n) ______ is an example of informal, written communication.
   A. engineering notebook
   B. grant proposal
   C. journal article
   D. technical report

______ 4. Which of the following is *not* a practice that should be followed when keeping an engineering notebook?
   A. Always use a pencil or marker.
   B. Cross out any remaining space on a page so work cannot be inserted after the date the page was created.
   C. If you make a mistake, cross it off, and add the corrections, along with your initials.
   D. Place your signature and the date at the bottom of every page.

______________________ 5. ______ writing is a formal type of expository writing, which is a form of written communication used to explain, describe, or inform.

______________________ 6. *True or False?* The subject of the sentence performs the action in active voice.

______________ 7. Technical reports, progress reports, specification sheets, and proposals are all examples of technical ______.

8. In a technical report, what information is typically provided on the *cover page*?

____

____

____

____

______ 9. Which of the following is found in the body of a technical report?
A. The abstract
B. The introduction
C. Lists of tables and figures
D. References

10. In a technical report, what does the *problem statement* describe?

____

____

____

____

______________ 11. *True or False?* The results section of a technical report provides an interpretation of the results of testing the solution.

______ 12. A(n) ______ provides only the information necessary to broadly describe a project; it does not include too much detail.
A. executive summary
B. journal article
C. status report
D. technical report

13. What is typically included in a *design proposal*?

____

____

____

____

______ 14. In which of the following types of technical documents is it acceptable to use informal writing conventions?
A. Emails
B. Social media
C. Traditional letters
D. None of the above.

Match each section of a technical report to its corresponding description.

A. Abstract
B. Appendices
C. Background
D. Introduction
E. Methods and procedures
F. Objective
G. Table of contents

_______ 15. A short summary of the whole project, stating what was done, how it was done, and what the results were

_______ 16. A list of all the major sections of the report, along with the beginning page number of each

_______ 17. A broad overview of the project that focuses on the purpose of the project but does not include an explanation of how the project was completed or any final results or conclusions

_______ 18. A section that defines the results that the project aims to achieve in solving the problem and does not include the work that has been done or a detailed work plan

_______ 19. A section that provides a detailed explanation of what is already known in relation to the stated problem and the information needed to begin thinking about how the problem could be solved

_______ 20. A section that focuses on the ways in which the solution was developed and evaluated but does not include the results

_______ 21. A section that includes any additional information needed to understand what is written in the body of the report, such as engineering drawings, mathematical analyses, illustrations, or photographs

22. Briefly explain *empirical research.*

____

____

____

______________________ 23. A scholarly paper that presents an extensive explanation of the current knowledge related to a topic based on a thorough examination of related written works is called a(n) _______.

______________________ 24. A(n) _______ is a sum of money given by the government, a private foundation, or a public corporation for a specific purpose.

_______ 25. Which of the following types of information are *not* generally included on a bill of materials?

A. Cost information for each part
B. The material each part is made of
C. The quantity of each part needed
D. The size of each part

______________________ 26. *True or False?* Specification sheets communicate the important properties a material must possess for a specific application.

27. Name five types of properties that might be provided on a specification sheet.

____

____

____

______ 28. Manufacturers prepare one type of technical data sheet to communicate the specifications for products they have on the market, called ______ materials and components.
A. nonstandard
B. on-the-shelf
C. optical
D. standard

____________________ 29. *True or False?* Engineering drawings are used to communicate a design solution idea or the information needed to produce the design solution.

30. Give one example of informal oral communication and one example of formal oral communication.

____

____

______ 31. Which of the following types of presentations is presented to potential stakeholders as a means to justify a potential project?
A. Critical design review
B. Final design presentation
C. Preliminary design review
D. Problem proposal

____________________ 32. The purpose of a(n) ______ design review is to present a developed solution pathway for solving a proposed problem to a panel of team members and experts.

33. Briefly describe what should be included in a final design presentation.

____

____

____

____________________ 34. It is important to tailor all presentations to your ______ by adjusting the terminology and amount of detailed information provided.

____________________ 35. *True or False?* When using presentation software, various slide designs, fonts, and colors should be used throughout the presentation.

_______ 36. When delivering a presentation, which of the following is *not* a good practice to follow?

A. Practice delivering the presentation in the mirror, as well as in front of friends and family, prior to delivering the actual presentation.
B. Read from your presentation slides, or memorize your script.
C. Speak loudly enough so your entire audience can hear you clearly.
D. Tailor your presentation to the audience.

______________________ 37. *True or False?* When delivering a presentation, you should answer questions honestly and professionally, admitting when you do not know the answer.

Name ______________________ Date __________ Class ______________

CHAPTER 9

Technology as a System

Complete the following questions and problems after carefully reading the corresponding textbook chapter.

____________ 1. Any entity or object that consists of parts, each of which has a relationship with all other parts and the entity as a whole, is called a(n) _____.

2. Give four examples of a natural system.

____

____

____

_____ 3. The process of understanding and mentally exploring systems in multiple ways is called systems _____.
A. modeling
B. society
C. technology
D. thinking

Match the components of a technological system to the corresponding letters in the following figure.

Universal Systems Model

A
B
C
D

Goodheart-Willcox Publisher

_____ 4. Feedback

_____ 5. Inputs

_____ 6. Outputs

_____ 7. Processes

_____________ 8. *True or False?* Inputs are the resources that go into a system and are used by the system.

9. Name four basic inputs to technological systems.

____

____

____

____

_____ 10. Which of the following bring specific knowledge, attitudes, and skills to a technological system?
A. Finances
B. Materials
C. People
D. Tools and machines

11. What is *information*?

____

____

____

____

_____ 12. _____ are the money and credit necessary for the economic system to operate.
A. Finances
B. Materials
C. People
D. Tools and machines

_____________ 13. _____ is a measurement of how long an event lasts.

_____________ 14. *True or False?* The procedure used to develop technological products and systems is called the scientific method.

_____ 15. In which of the following steps of the engineering design process does a person or group develop basic information about the problem and the design limitations?
A. *Develop a Design Solution*
B. *Gather Information*
C. *Identify and Define a Problem*
D. *Model and Make a Solution*

_____________ 16. During the engineering design process, a person or group obtains the knowledge necessary to develop a solution design by conducting research and studying _____.

_____________ 17. *True or False?* In the *Model and Make a Solution* step of the engineering design process, a person or group develops and refines several possible solutions.

18. Explain what happens during the *Evaluate the Solution* step of the engineering design process.

____

____

____

_______ 19. In which of the following steps of the engineering design process does a person or group share the results of the process by preparing the documents and presentations needed to share the evaluation outcomes?
A. *Communicate the Final Solution*
B. *Develop a Design Solution*
C. *Evaluate the Solution*
D. *Model and Make a Solution*

_______________ 20. *True or False?* Production processes are actions completed to perform the function of a technological system.

21. What are *management processes*?

____

____

____

Match the management activities to the corresponding letters in the following figure.

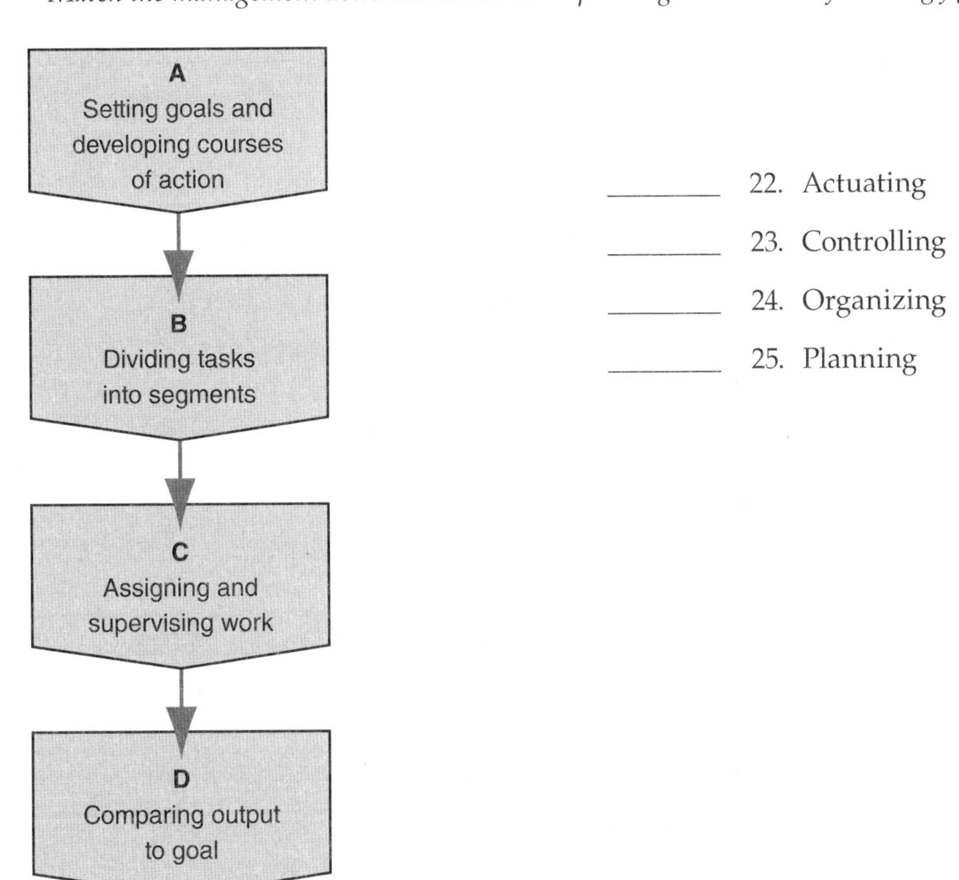

Goodheart-Willcox Publisher

_______ 22. Actuating

_______ 23. Controlling

_______ 24. Organizing

_______ 25. Planning

_______ 26. Which of the following is the process of using information about the outputs of a system to regulate the inputs?
A. Control
B. Feedback
C. Management
D. Production

_______________________ 27. The feedback loop that causes management and production activities to change through evaluation, feedback, and corrective action is called _______.

Name ______________________ Date __________ Class ______________

CHAPTER 10

Inputs to Technological Systems

Complete the following questions and problems after carefully reading the corresponding textbook chapter.

1. Name four categories of inputs to technological systems.

____

____

Match each type of person to its description.

A. Consumer
B. Entrepreneur
C. Manager
D. Mechanic
E. Production worker
F. Support staff
G. Technician

_______ 2. A person who processes materials, builds structures, operates transportation vehicles, services products, or produces and delivers communication products

_______ 3. A skilled worker in a laboratory or product-testing facility

_______ 4. A skilled worker in a service operation

_______ 5. An individual who has a vision and accepts the financial risks of creating a small business

_______ 6. A person who organizes and directs the work of others in a business

_______ 7. A person who carries out tasks such as keeping financial records, maintaining sales documents, and developing personnel systems

_______ 8. A person who financially supports technological systems by spending money on products or services

_______ 9. Which of the following is the ability to see a need or way to make life easier and design systems and products to meet the need or desire?
A. Biotechnology
B. Creativity
C. Knowledge
D. A partnership

_______ 10. The artifacts that expand what humans are able to do are called _______.

Match the types of hand tools to the corresponding letters in the following figure.

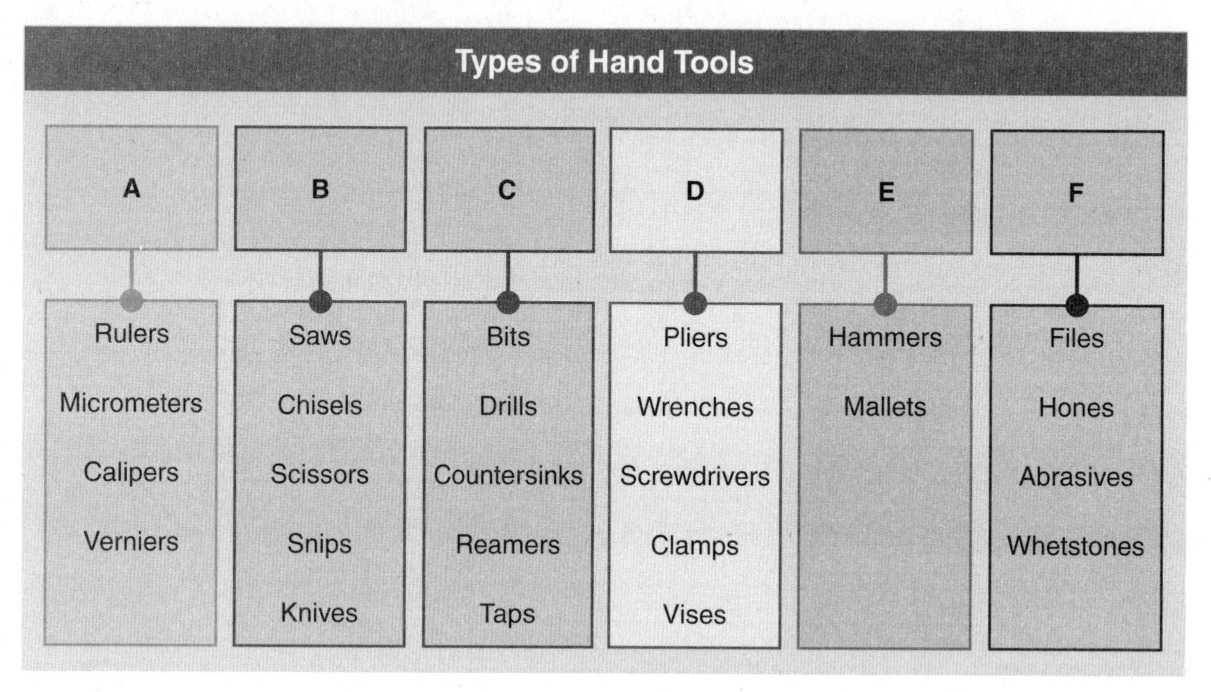

Goodheart-Willcox Publisher

_______ 11. Cutting
_______ 12. Drilling
_______ 13. Gripping
_______ 14. Measuring
_______ 15. Polishing
_______ 16. Pounding

_______ 17. A(n) _______ is an artifact that transmits or changes the application of power, force, or motion.
A. exhaustible resource
B. hand tool
C. machine
D. natural resource

_______ 18. *True or False?* An inclined plane is a simple machine that multiplies the force applied to it and changes the direction of a linear force.

19. What is a *lever arm*?

____

____

____

______________________ 20. The fulcrum is between the load and the effort in ______-class levers.

______________________ 21. *True or False?* In third-class levers, the load is between the effort and the fulcrum.

22. In which class of levers is the effort placed between the load and the fulcrum?

____

23. What is a *distance multiplier*?

____

____

____

______ 24. Which of the following simple machines is a shaft attached to a disk?
A. A lever
B. A pulley
C. A screw
D. A wheel and axle

______________________ 25. Grooved wheels attached to an axle are called a(n) ______.

26. What is a *wedge*?

____

____

____

______ 27. Which of the following simple machines is an inclined plane wrapped around a shaft?
A. A pulley
B. A screw
C. A wedge
D. A wheel and axle

______________________ 28. All ______ is made up of one or more of the elements occurring naturally on the earth.

29. What is a *natural resource*?

____

____

____

____

_______ 30. Which of the following is *not* an example of a natural material?
A. Carbon
B. Iron
C. Petroleum
D. Plastic

_______ 31. A material that is manufactured is called a(n) _______ material.

_______________ 32. Materials that come from living organisms are called _______ materials.
A. composite
B. inorganic
C. natural
D. organic

_______________ 33. Materials that do *not* come from living organisms are called _______ materials.

34. What is an *exhaustible resource*?

____

____

____

_______ 35. _______ materials are produced by living things.
A. Genetic
B. Inexhaustible
C. Inorganic
D. Natural

_______________ 36. _______ are materials that easily disperse and expand to fill any space.

37. What is a *liquid*?

____

____

____

38. List four common categories of properties of materials.

____

____

_______ 39. All raw facts and figures collected by people and machines are known as ______.
A. data
B. scientific information
C. technological information
D. knowledge

40. Name two areas that information can be grouped into.

____

____

_______________________ 41. *True or False?* Technological information is organized data about the laws and natural phenomena in the universe.

_______ 42. ______ is organized data about the design, production, operation, maintenance, and service of human-made products and structures.
A. Humanities information
B. Knowledge
C. Scientific information
D. Technological information

_______________________ 43. Organized data about the values and actions of individuals and society is ______ information.

44. Name four major types of energy.

____

____

_______________________ 45. *True or False?* Radiant energy is stored in a substance and released by chemical reactions.

_______________________ 46. Thermal energy comes from the increased molecular action that ______ causes.

_______ 47. The sun, fire, and other matter, including light, radio waves, x-rays, and UV and infrared waves, produce ______ energy.
A. electrical
B. nuclear
C. radiant
D. thermal

_______ 48. When money is raised by selling partial ownership in a technological system or company, it is called ______ financing.
A. debt
B. equity
C. first-class
D. time

_______________________ 49. A(n) ______ resource cannot be entirely used up or consumed.

______ 50. Which of the following statements is *not* true regarding a corporation?
A. It is a legal entity formed by people to own a company.
B. It is owned by one person.
C. It sells shares of the company.
D. Shares of the corporation are certificates of ownership.

51. What is *debt financing*?

____

____

____

____

______ 52. Technology has accelerated the use of and changed our standards of measurement for ______.
A. energy
B. finances
C. information
D. time

Name ______________________________ Date __________ Class ____________________

CHAPTER 11

Technological Processes

Complete the following questions and problems after carefully reading the corresponding textbook chapter.

______ 1. Which of the following are used to develop and operate technological systems?
A. Management processes
B. Problem-solving and engineering design processes
C. Production processes
D. All of the above.

____________________ 2. *True or False?* The trial-and-error method of developing a solution follows the same steps as the engineering design process.

3. List four of the steps in the engineering design process.

____

____

____

____

____________________ 4. In the engineering design process, once the problem is defined, designers seek solutions by gathering all potentially valuable ______, which is then carefully reviewed.

____________________ 5. *True or False?* Models are used to test product and structure designs.

______ 6. Throughout the engineering design process, engineers often test and ______ different ideas or parts of a design to determine whether or not these elements will perform as desired.
A. communicate
B. evaluate
C. gather
D. identify

________________ 7. Graphics, bills of materials, and material specifications are used to ______ the characteristics of a product, structure, media, or system.

______ 8. Which of the following is the first step in agricultural practices?
A. Conversion and processing
B. Growth
C. Harvesting
D. Propagation

________________ 9. The step in the agricultural process that involves removing edible parts of plants from trees and stock and butchering animals to produce meat and other consumable products is called ______.

________________ 10. *True or False?* Harvesting involves changing food products into foodstuffs.

11. What is *telecommunications technology*?

____

____

____

________________ 12. The manipulation of data to produce useful information is called ______.

13. List the steps involved in all communication technologies.

____

____

____

________________ 14. The process in which coded information is changed back into a recognizable form is called ______.

________________ 15. *True or False?* To receive information is to retain it for later use.

______ 16. A structure erected to protect people and machines from the outside environment is called a ______.
A. building
B. civil engineering structure
C. foundation
D. heavy engineering structure

________________ 17. A structure used for business purposes is called a(n) ______ structure.

18. List five steps included in most construction projects.

____

____

____

____

____

______________________ 19. *True or False?* A foundation is the base of a structure.

______________________ 20. The maintenance, repair, and reconditioning that structures require to maintain their integrity is called ______.

______________________ 21. *True or False?* Energy is the rate at which work is accomplished.

______ 22. The energy found in moving objects is ______ energy.
- A. chemical
- B. mechanical
- C. radiant
- D. thermal

______________________ 23. Energy contained in molecules of a substance is called ______ energy.

______________________ 24. *True or False?* Electrical energy is energy in the form of electromagnetic waves.

25. What is *radiant energy*?

____

____

____

______________________ 26. The energy associated with moving electrons is called ______ energy.

27. When gathering resources for energy and power processes, the type of gathering technique used depends on what?

____

____

______ 28. Solar panels ______ radiant energy into electrical energy.
- A. apply
- B. convert
- C. gather
- D. transmit

______________________ 29. *True or False?* Energy, in itself, is of little use to people until it is applied to a task.

Match the stages of manufacturing activities to the corresponding letters in the following figure.

Weyerhauser Co.

A

MeadWestavco

B

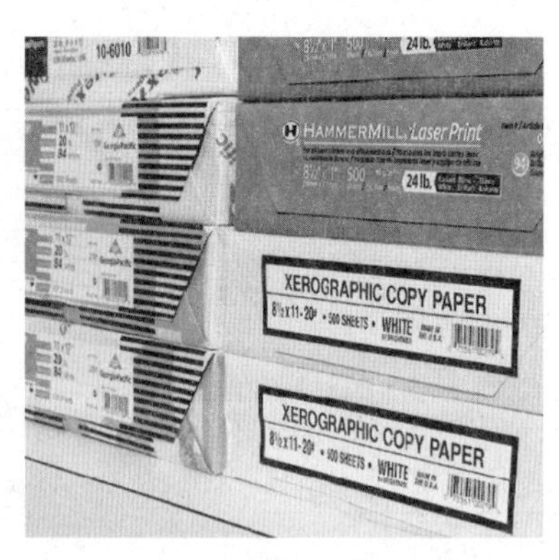

Goodheart-Willcox Publisher

C

_______ 30. Obtaining resources

_______ 31. Producing finished products

_______ 32. Producing industrial materials

_______________________ 33. The process of obtaining materials from the earth by pumping them through holes drilled into the earth is called ______.

Match each term to its corresponding description.

A. Condition
B. Form
C. Separate
D. Finish
E. Assemble

_______ 34. Coating surfaces of products to protect them, make them more appealing to consumers, or both

_______ 35. Apply force using a die or roll to reshape materials

_______ 36. Using tools to shear or machine away unwanted material

_______ 37. Using heat, chemicals, or mechanical forces to change the internal structure of a material

_______ 38. Bringing materials and parts together to finish product

39. Name three areas of medicine in which medical tools are used.

____

____

_______________________ 40. *True or False?* In medical processes, diagnosis is using knowledge, technological devices, and other means to determine the causes of abnormal body conditions.

______ 41. The movement of humans and their possessions from one place to another is called ______.
A. energy
B. guidance
C. propulsion
D. transportation

42. List four main transportation media.

____

____

____________________ 43. *True or False?* Vehicular systems are onboard technical systems that operate a vehicle.

______ 44. Which of the following vehicular systems generates motion through energy conversion and transmission?
A. Control
B. Guidance
C. Propulsion
D. Suspension

____________________ 45. The vehicular system that gathers and displays information so the vehicle can be kept on course is called ______.

46. What are the two types of support systems in transportation processes?

____

____

Match each term to its corresponding description.

A. Organizing
B. Controlling
C. Planning
D. Actuating

______ 47. Developing goals and objectives

______ 48. Structuring an activity, establishing procedures to reach a goal, and drawing lines and levels of authority within an enterprise

______ 49. Checking the outputs against the plan

______ 50. Starting actual work

Name ______________________ Date __________ Class ______________

CHAPTER 12

Outputs, Feedback, and Control

Complete the following questions and problems after carefully reading the corresponding textbook chapter.

______________ 1. *True or False?* Outputs can be categorized as desirable or undesirable, intended or unintended, and immediate or delayed.

______________ 2. Inputs are processed into ______.

______ 3. Outputs that benefit people are called ______ outputs.
A. desirable
B. immediate
C. intended
D. undesirable

______________ 4. Outputs that are *not* wanted or planned for are called ______ outputs.

______ 5. Outputs that were *not* anticipated when the system was designed are called ______ outputs.
A. delayed
B. intended
C. undesirable
D. unintended

______________ 6. Products or services designed for use now are called ______ outputs.

______________ 7. *True or False?* Outputs that occur at a later date than expected or planned are called unintended outputs.

8. What does feedback do with the information from an output of a system?

____

____

____

_______________ 9. A system that does not use output information to adjust the process is called a(n) ______ system.

10. What is *wage control*?

____

____

____

Match each term to its corresponding description.

A. Analytical system
B. Optical sensor
C. Thermal sensor
D. Judgmental system
E. Fluidic controller
F. Mechanical controller
G. Monitoring device
H. Electromechanical controller

______ 11. A device used to determine light level or changes in light intensity

______ 12. A device used to determine changes in temperature

______ 13. A sensor that gathers information about an action being controlled

______ 14. A data-comparing device that makes comparisons mathematically or scientifically

______ 15. A data-comparing device that allows human opinions and values to enter into the control process

______ 16. A type of adjusting device that uses cams, levers, and other types of linkages to adjust machines or other devices

______ 17. A type of adjusting device that uses electromagnetic coils and forces to move control linkages and operate switches to adjust machines or other devices

______ 18. A type of adjusting device that uses fluids to adjust machines or other devices

______ 19. Which of the following are technological control systems that require humans to adjust the processes?

A. Automatic control systems
B. Closed-loop control systems
C. External control systems
D. Manual control systems

_______________ 20. A technological control system that can monitor, compare, and adjust systems without human interference is called a(n) ______ control system.

_______________ 21. *True or False?* A good example of a manual control system is the automobile.

______ 22. A ______ is a good example of automatic control.
A. bicycle
B. camera
C. gas stove
D. thermostat

______________________ 23. Controversies, such as those concerning the types of nets that should be used in fishing, are examples of ______ control.

Name ______________________ Date __________ Class ______________

CHAPTER 13

Designing the World through Engineering

Complete the following questions and problems after carefully reading the corresponding textbook chapter.

______________ 1. The ______ world encompasses anything created by humans that is *not* represented in the ______ world.

______ 2. Which of the following are *not* part of the natural world?
A. Animals
B. Plants
C. Rocks
D. Tools

______________ 3. ______ design is associated with achieving specific goals and the application of mathematical and scientific knowledge.

4. Name four categories of technologies.

____

____

____

____

______________ 5. ______ technologies cover a range of processes, systems, and procedures to create products people need or want.

______________ 6. *True or False?* Energy and power technologies are technologies that move people and products from one location to another through land, water, air, and space travel.

Put the stages of a building's life cycle in proper order.

A. Construction
B. Demolition
C. Operation
D. Proposal
E. Renewal
F. Upkeep

__D__ 7. Stage 1

__A__ 8. Stage 2

__C__ 9. Stage 3

__F__ 10. Stage 4

__E__ 11. Stage 5

__B__ 12. Stage 6

__Energy__ 13. ______ is the ability to do work.

__B__ 14. ______ is the amount of energy consumed over time by a technological system.
A. Manufacturing
B. Power
C. Transportation
D. Work

__Information and Communication__ 15. Technologies that provide the ability to record, store, manipulate, analyze, and transmit data across various modes are called ______ and ______ technologies.

__Medical__ 16. ______ technologies create the tools used in the prevention, diagnosis, monitoring, and treatment of illness, as well as in the repair of injury.

__True__ 17. *True or False?* Biomedical engineers apply principles of engineering to design new technologies.

18. What is *biotechnology*?

The application of living organisms, in whole or in part, to create technology.

Name ______________________________ Date __________ Class ____________________

CHAPTER 14

Processing Resources

Complete the following questions and problems after carefully reading the corresponding textbook chapter.

1. What is *form utility*?

____

____

____

______ 2. What do we call the production activities that produce structures?
A. Construction
B. Mass production
C. Primary manufacturing
D. Secondary manufacturing

____________________ 3. The process of efficiently creating a large number of standardized products by separating the assembling process into small and easily managed steps is called ______ production.

4. Briefly explain *just-in-time (JIT) manufacturing*.

____

____

____

____

Match the types of materials and products to the corresponding letters in the following figure.

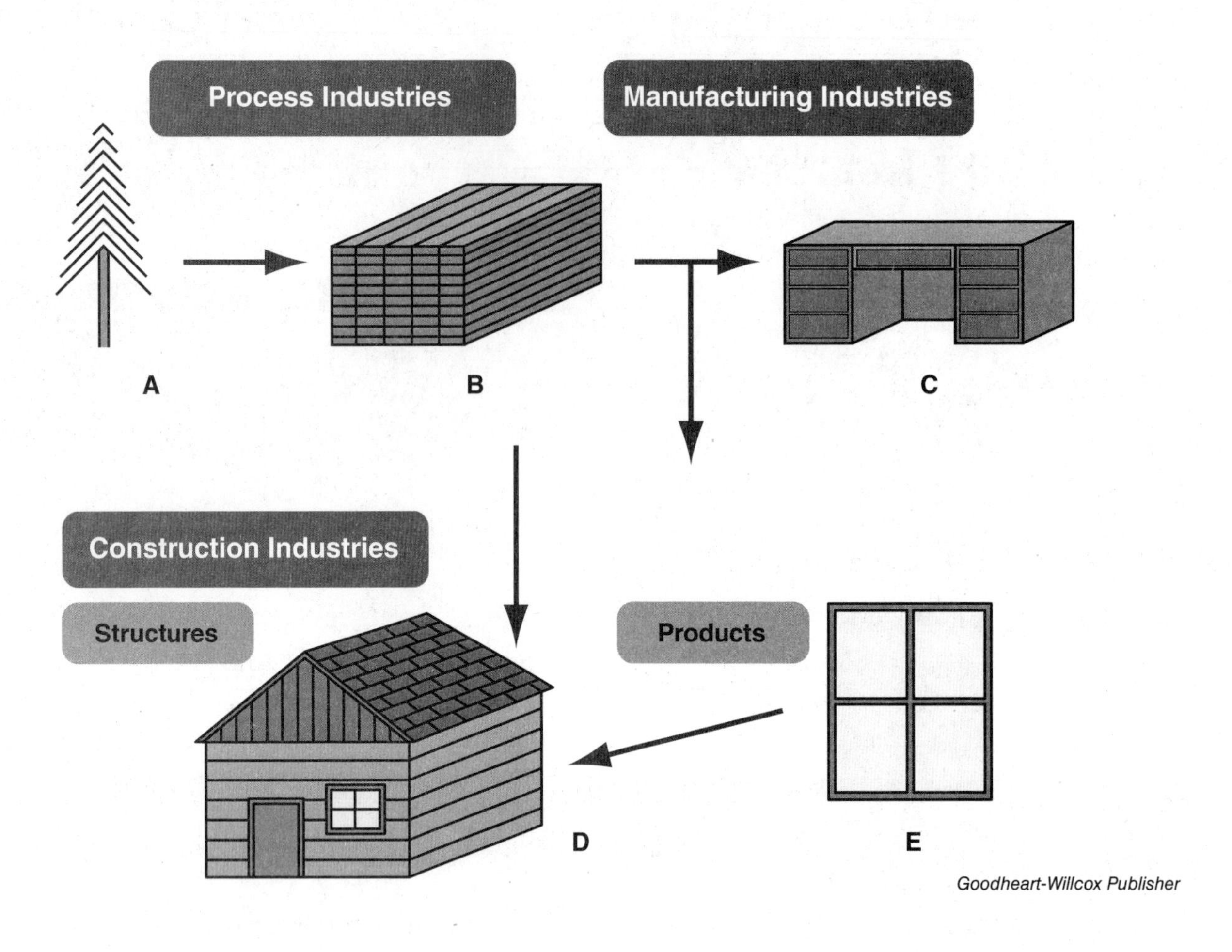

Goodheart-Willcox Publisher

_______ 5. Constructed building

_______ 6. Consumer product

_______ 7. Industrial material

_______ 8. Industrial product

_______ 9. Natural resource

_______ 10. The English translation of *Kaizen* is _______.
- A. change for the worse
- B. good change
- C. lean manufacturing
- D. stabilization

_______________________ 11. Systems that move materials, parts, or products in a fixed path along a production line are called _______ systems.

_______________________ 12. *True or False?* Mass customization is the ability of manufacturing systems to produce standardized products, though often at a high cost.

13. What is *automation*?

____

____

____

_______ 14. Which of the following are mechanical devices that can perform tasks automatically or with varying degrees of direct human control?
   A. Automated guided vehicles
   B. Computer numerical control (CNC) machines
   C. Programmable logic controllers (PLCs)
   D. Robots

_______________________ 15. *True or False?* The first industrial robot, developed in 1962, was a pick-and-place robot.

16. Briefly describe *CNC*.

____

____

____

____

_______ 17. Which of the following is a device that uses microprocessors to control machines or processes?
   A. An automated guided vehicle
   B. A CNC machine
   C. A PLC
   D. A robot

_______________________ 18. The intelligence exhibited by a manufactured device or system is called _______ intelligence.

_______ 19. Which of the following is the fabrication of a scaled part or model of a product using three-dimensional (3-D) computer-aided design (CAD) data?
   A. Artificial intelligence (AI)
   B. CNC
   C. Rapid prototyping
   D. Subtractive manufacturing

20. What is the difference between additive manufacturing and subtractive manufacturing?

____

____

____

____

____

______ 21. The resource processing systems of manufacturing are called ______ manufacturing.
A. additive
B. primary
C. secondary
D. subtractive

______________________ 22. Material resources obtained during the normal life cycles of plants or animals through farming, fishing, and forestry are called _______ materials.

______________________ 23. *True or False?* The appearance of animal life is called birth.

______ 24. Which of the following is *not* a logging method?
A. Clear-cutting
B. Seed-tree cutting
C. Selective cutting
D. Timber cruising

______ 25. Which of the following is the process of gathering logs in a central location?
A. Bucking
B. Felling
C. Hauling
D. Yarding

______________________ 26. Material resources formed from the remains of prehistoric living organisms are called ______.

______________________ 27. *True or False?* Hydrocarbons are mixtures of carbon and hydrogen.

28. Name two types of fossil fuel resources.

____

____

______ 29. ______ is a combustible fossil fuel resource that occurs in porous rock and is composed of light hydrocarbons.
A. Coal
B. Lignite
C. Natural gas
D. Petroleum

______________________ 30. Bituminous coal is sometimes called ______ coal because it can be easily broken into various sizes.

______________________ 31. *True or False?* Anthracite coal has the lowest carbon content of all the types of coal.

32. What is a *seismographic study?*

____

____

____

____

____

______________________ 33. The mixture of water, clay, and chemicals that is pumped down a drill pipe is called ______.

______________________ 34. *True or False?* Horizontal drilling is drilling a well along a curve to reach petroleum or natural gas deposits that cannot otherwise be tapped.

35. Give a brief explanation of *hydraulic fracturing.*

____

____

____

____

Match the types of underground mining to the corresponding letters in the following figure.

_______ 36. Drift mining

_______ 37. Shaft mining

_______ 38. Slope mining

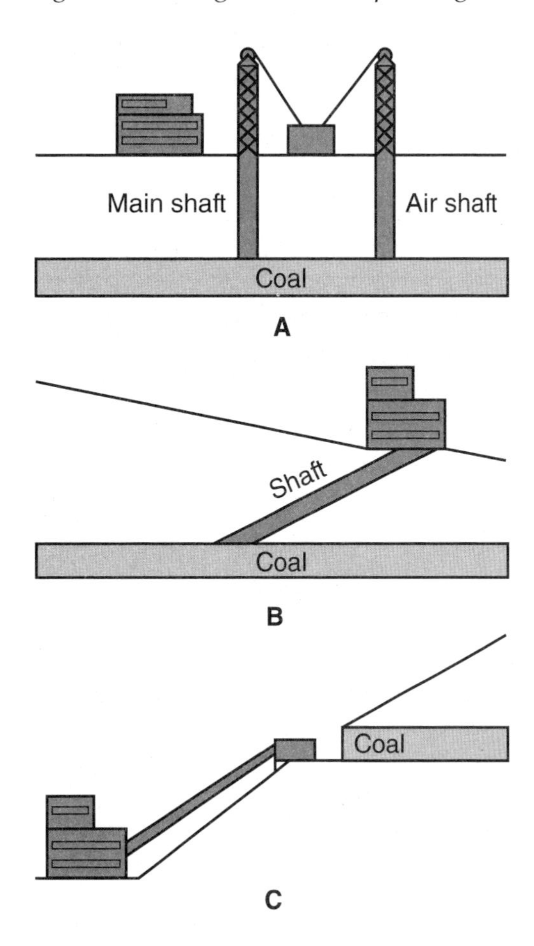

Goodheart-Willcox Publisher

39. List three groups of mineral families that have similar features.

____

____

40. Briefly explain *fluid mining.*

____

____

____

____

41. List two types of primary processes.

____

____

42. Name two types of lumber.

____

____

______________ 43. *True or False?* An engineered wood product is a composite material made from lumber production process waste materials and adhesives.

Identify the type of core used in each of the following illustrations.

_______ 44. Lumber core

_______ 45. Veneer core

_______ 46. Particleboard core

A

B

C

Goodheart-Willcox Publisher

47. What is a *thermal process*?

____

____

_______ 48. Which of the following statements regarding chemical and electrochemical processes is *true*?

A. They break down or build up materials by changing their chemical compositions.
B. They are used to produce synthetic fibers, pharmaceuticals, and other valuable products.
C. They are used to chemically combine different materials, such as zinc and steel, to prevent corrosion or rusting.
D. All of the above.

Name ______________________ Date __________ Class ______________

CHAPTER 15

Producing Products

Complete the following questions and problems after carefully reading the corresponding textbook chapter.

______________ 1. The actions used to change industrial materials into products are called _______ manufacturing processes.

2. List four groups of secondary manufacturing processes.

____

____

______ 3. A secondary manufacturing process that gives materials shape by introducing a liquid material into a mold is called a(n) _______ and molding process.
A. casting
B. conditioning
C. forming
D. separating

______________ 4. *True or False?* Liquids include molten and fluid materials.

______________ 5. _______ is a secondary manufacturing process that removes excess material to make an object of the correct size and shape.
A. Casting
B. Conditioning
C. Forming
D. Separating

6. What is *machining*?

____

____

____

Match the terms to the corresponding letters in the figure below.

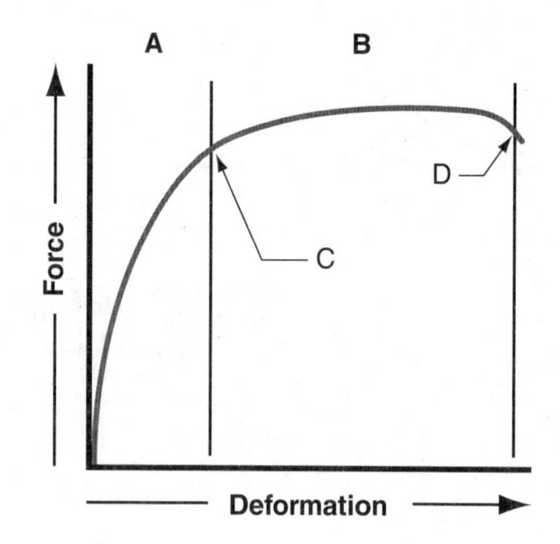

Goodheart-Willcox Publisher

_____ 7. Elastic range

_____ 8. Fracture point

_____ 9. Plastic range

_____ 10. Yield point

_______________ 11. _____ is a separating process in which opposing edges of blades, knives, or dies are used to fracture the unwanted material away from the work.

_____ 12. A secondary manufacturing process that changes the internal properties of a material is the _____ process.

A. casting
B. conditioning
C. forming
D. separating

_______________ 13. The use of _____ forces to change the internal structure of a material is mechanical conditioning.

_______________ 14. *True or False?* In thermal conditioning, chemical actions change the properties of a material.

_____ 15. A thermal conditioning process used on metals, in which the metal is heated and allowed to cool slowly, is called heat _____.

A. conditioning
B. drying
C. firing
D. treating

_______________ 16. An assembly process that uses mechanical forces, such as friction, to hold parts together is mechanical _____.

_______________ 17. *True or False?* Bonding holds parts together using mechanical forces.

18. Describe the difference between a press fit and a seam.

____

____

____

____

____

____

_____ 19. Which type of finish chemically changes the surface of products?
A. Converted surface finish
B. Paint
C. Varnish
D. None of the above.

______________ 20. A production system that individually produces unique products tailored to meet the specific needs of the customer is a(n) _____ production system.

______________ 21. *True or False?* A mass customization production system produces a large quantity of similar products in the least amount of time possible without interruption.

22. Briefly explain *batch production*.

____

____

____

_____ 23. The engineering design process is used to devise and refine a marketable and profitable _____.
A. input
B. material
C. product
D. resource

______________ 24. Just as societal and economic factors impact product _____ and development, product production impacts society and the economy.

25. What is a *durable product*?

____

____

____

_____ 26. A(n) _____ is designed to be used for only a short period of time.
A. continuous product
B. durable product
C. nondurable product
D. obsolete product

_____ 27. A(n) _____ is a situation where an action diminishes one aspect of something in order to enhance another aspect.

_____ 28. A diagram that uses symbols to represent a sequence of actions or operations of a complex system, such as a manufacturing facility, is called a(n) _____.

_____ 29. *True or False?* A fixture is a device that guides machines and tools performing manufacturing operations.

30. Briefly explain *machine vision*.

_____

_____

_____

_____

Name Jessica Pucci Date 1/21/22 Class

CHAPTER 16

Meeting Needs through Materials Science and Engineering

Complete the following questions and problems after carefully reading the corresponding textbook chapter.

True 1. *True or False?* The study of solid materials at the atomic level is called materials science and engineering.

2. How does materials science and engineering classify solid materials?...

They're classified by their atomic and molecular characteristics and differences in properties.

D 3. Which of the following terms refers to the process of extracting metal from an ore through heating and cooling?

A. Bonding
B. Conducting
C. Insulating
D. Smelting

False 4. *True or False?* An alloy is a pure metal.

B 5. Which of the following is *not* a typical division of materials?

A. Ceramics
B. Irons
C. Metals
D. Polymers

Match each time period to its defining characteristic.

A. Advanced materials age
B. Bronze Age
C. Iron and Steel Age
D. Nonferrous and Polymer Age
E. Stone Age

E 6. The manipulation of natural materials to create tools for survival

B 7. The introduction of smelting to create tools for hunting, defense, and farming

C 8. Materials shaped increasingly by blacksmiths; smelting in bloomery furnaces

D 9. The introduction of materials *not* containing iron, especially synthetic plastic

A 10. The creation of new, stronger, and more efficient materials

Match the components of an atom to the corresponding letters in the figure.

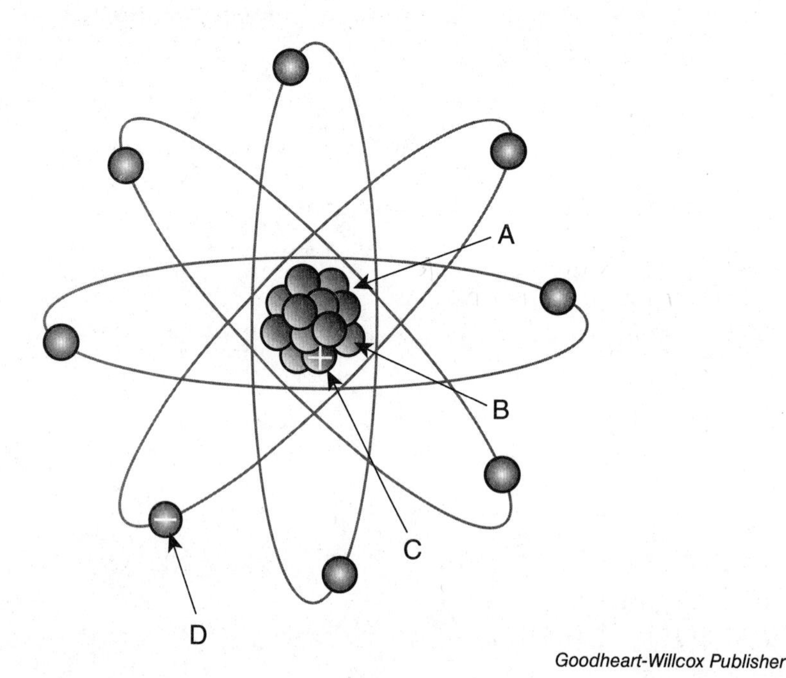

Goodheart-Willcox Publisher

D 11. Electron

B 12. Neutron

A 13. Nucleus

C 14. Proton

Valance electrons 15. Electrons in the outermost shell of an atom are called _______ electrons.

False 16. *True or False?* An ion is a balanced atom.

17. What is a *conductor*?

A material that contains atoms with only one valance electron that is loosely bound and can become easily free to conduct electrical current

B 18. _______ do not conduct electrical current under typical conditions, due to valence electrons that are tightly bound to the atoms.
A. Elastomers
B. Insulators
C. Polymers
D. Semiconductors

Elements 19. Substances that cannot be broken down any further through chemical reactions are called _______.

20. What is an *atomic number*?

The number of protons in the nucleus of their atoms, which is the same as the number of electrons that are distributed through electron orbitals.

C 21. Areas within an atom where electrons have a high potential of being located are electron _______.
A. bonds
B. molecules
C. orbitals
D. structures

chemical bonding 22. When the valence electrons of atoms are attracted to one another and transferred or shared between the atoms, _______ occurs.

True 23. *True or False?* The bonding of two or more atoms creates molecules.

24. What is *molecular structure*?

The structure and connections between molecules and atoms.

C 25. Which of the following is *not* a type of material properties?
A. Chemical properties
B. Mechanical properties
C. Metallic properties
D. Physical properties

False 26. *True or False?* Mechanical properties describe a material's characteristics as a result of a chemical reaction or change.

Flammability 27. The ease with which a material will ignite or burn is called ______.

C 28. ______ is the maximum pulling force a material can withstand, prior to failure.
A. Ductility
B. Hardness
C. Tensile strength
D. Thermal fatigue

29. What are *physical properties?*
The characteristics due to the structure of a material, including size, shape, density, moisture content, and porosity.

C 30. ______ is the extent to which a material can transfer heat.
A. Ductility
B. Tensile strength
C. Thermal conductivity
D. Thermal fatigue

31. What are *acoustical properties?*
They describe how a material reacts to sound waves.

metals 32. Materials that have strong electrical and thermal conductivity; are solid at normal temperatures; are shiny, hard, malleable, and ductile; can be difficult to burn; and have a high density are ______.

true 33. *True or False?* Metals containing a single type of atom are called pure metals.

34. What is an *alloy?*
Metals containing multiple atoms or a mixture of metals.

A 35. Which of the following is *not* a class of polymer?
A. Ceramics
B. Elastomers
C. Thermosplastics
D. Thermosets

Thermoplastics 36. Polymers that do not share electrons between atoms, and thus have no covalent bonding, are called ______.

37. What is an *elastomer*?

They are elastic polymers that can stretch, but return to their original shape when released.

A 38. ______ are materials characterized by high brittleness, poor electrical and thermal conductivity, very high melting points, and nonflammability.

A. Ceramics
B. Composites
C. Metals
D. Polymers

True 39. *True or False?* A material that is between an insulator and a conductor, in terms of its ability to conduct electrical current, is a semiconductor.

40. What is an *advanced material*?

Innovative materials to meet the needs of high-tech applications.

D 41. Materials that have properties that can be modified in a controlled manner by external forces are ______ materials.

A. composite
B. nanostructured
C. semiconductor
D. smart

Name ________________________ Date __________ Class ________________

CHAPTER 17

Constructing Structures

Complete the following questions and problems after carefully reading the corresponding textbook chapter.

__________________ 1. *True or False?* The three main types of buildings are residential, commercial, and heavy engineering.

2. Name two activities people do in buildings.

____

____

______ 3. A(n) ______ building is a structure used for business purposes.
A. commercial
B. cultural
C. industrial
D. residential

__________________ 4. A structure that houses manufacturing processes is a(n) ______.

5. What is the purpose of a monument?

____

____

____

______ 6. City halls, post offices, police stations, firehouses, state capitols, and courthouses are examples of ______ buildings.
A. agricultural
B. commercial
C. cultural
D. government

__________________ 7. Transportation ______ are used to load and unload passengers and cargo from transportation vehicles.

______ 8. Which type of building is built in a factory, usually in two halves?
A. Commercial building
B. Industrial building
C. Manufactured home
D. Transportation terminal

______________ 9. *True or False?* Zoning laws, building codes, and professional practices put constraints on building plans.

______ 10. ______ are regulations controlling the design and construction of a structure to provide for human safety and welfare.
A. Budgets
B. Building codes
C. Best practices
D. Zoning laws

11. List four of the basic steps of most construction projects.

____

____

____

____

Match the terms to the correct letters in the following figure.

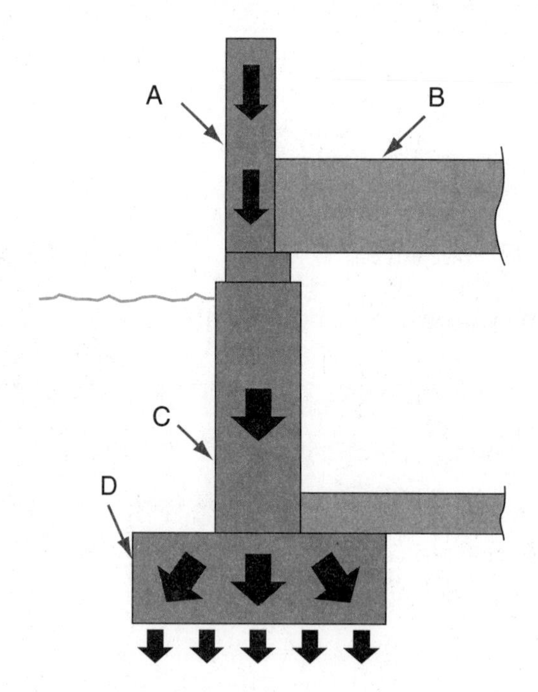

Goodheart-Willcox Publisher

______ 12. Floor joist

______ 13. Footing

______ 14. Foundation wall

______ 15. Wall

Match the terms to the correct letters in the following figure.

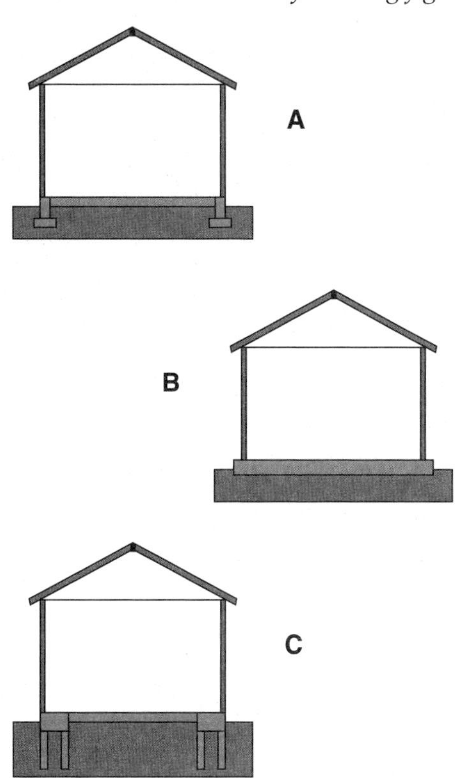

Goodheart-Willcox Publisher

______ 16. Pile foundation

______ 17. Slab foundation

______ 18. Spread foundation

______________________ 19. Concrete that has wire mesh or steel bars embedded into it to increase its tensile strength is called ______ concrete.

______________________ 20. *True or False?* A wood piece bolted to the top of a foundation is called the subfloor.

21. What is a *floor joist*?

____

____

____

______ 22. A ______ is the strip at the bottom of a framed wall.
- A. header
- B. sole plate
- C. stud
- D. top plate

______ 23. Which of the following hold up headers?
- A. Cripple studs
- B. Sole plates
- C. Top plates
- D. Trimmer studs

24. What is a *rafter*?

____

____

____

Match the terms to the corresponding letters in the following figure.

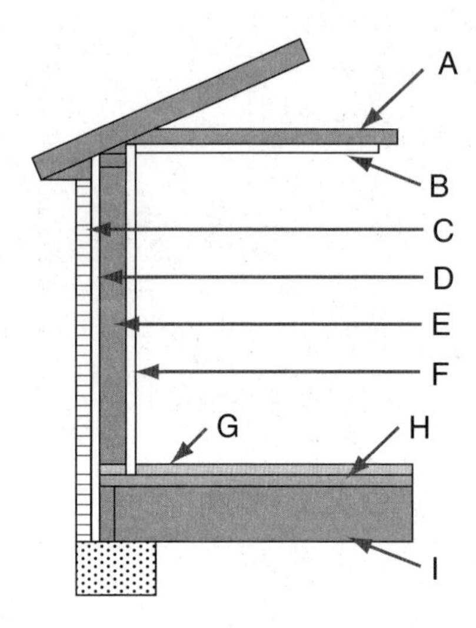

Goodheart-Willcox Publisher

_______ 25. Brick

_______ 26. Ceiling

_______ 27. Ceiling joist

_______ 28. Drywall

_______ 29. Floor joist

_______ 30. Flooring

_______ 31. Sheathing

_______ 32. Stud

_______ 33. Subfloor

_________________ 34. *True or False?* A shingle is a board used to finish the ends of the rafters and the overhang.

35. What is a *soffit*?

____

____

____

______ 36. Which of the following is *not* a major utility system?
A. Climate control
B. Communications
C. Plumbing
D. Transportation

____________________ 37. *True or False?* Electrical systems deliver electrical power to houses.

38. What is *wastewater*?

____

____

____

____________________ 39. A type of indirect heating method that heats air as a conduction medium is a(n) ______-air heating system.

____________________ 40. *True or False?* Hot water heating is a type of direct heating method that uses water to carry heat.

41. Name two types of communication systems.

____

____

______ 42. Trees, shrubs, and grass that are planted to help prevent erosion and improve the appearance of a site are called ______.
A. earth dams
B. footing
C. landscaping
D. traps

____________________ 43. Construction activities can be used to produce civil structures or ______ structures.

______ 44. A(n) ______ bridge uses small triangular parts to support the deck.
A. arch
B. beam
C. suspension
D. truss

45. What is a *suspension bridge*?

____

____

____

______________________ 46. *True or False?* Most telecommunication technology relies on constructed towers to support antennas.

______ 47. A(n) ______ dam has a vertical upstream wall and a sloping downstream wall.

A. arched
B. buttress
C. gravity
D. rock

______________________ 48. A type of dam that has a solid upstream side and is supported on the downstream side with a series of supports is a(n) ______ dam.

Name ______________________________ Date __________ Class ____________________

CHAPTER 18 Meeting Needs through Architecture and Civil Engineering

Complete the following questions and problems after carefully reading the corresponding textbook chapter.

______________________ 1. Planning and designing the appearance of structures evolved into the professional practices of civil engineering and ______.

______________________ 2. *True or False?* Advancements in materials and construction methods and complexities in building structures required more scientific and technical knowledge, resulting in collaboration between professionals in architecture and professionals in engineering.

3. What is *architecture*?

____

____

____

____

______________________ 4. *True or False?* Geotechnical engineering is the study of the framework of structures: designing, analyzing, and constructing structural components or assemblies to resist the stresses and strains of loads and forces that affect them.

5. What is *solid mechanics*?

____

____

____

____

6. What is *transportation engineering*?

____

____

____

____

______________ 7. ______ engineering is managing the use of natural resources to minimize the negative impacts that human activity can have on the environment.

______ 8. A(n) ______ is the top view of a building that includes cross-sectional views of different levels.
A. bubble plan
B. elevation
C. orthographic projection
D. plan view

______ 9. A(n) ______ is a rough drawing used when brainstorming a structure's layout.
A. bubble plan
B. elevation
C. orthographic projection
D. plan view

______ 10. ______ is a pushing or pulling applied to an object as a result of an interaction with another object.
A. A force
B. A load
C. Strain
D. Stress

11. What is *strain*?

____

____

____

______ 12. Which type of force pulls material apart?
A. Bending stress
B. Compression
C. Tension
D. Torsion

______________ 13. The force that pushes on material and squeezes it is called ______.

______________ 14. *True or False?* Shear causes materials to experience both tension and compression.

15. What is *torsion*?

____

____

____

____

______________________ 16. ______ acting on a structure originate from a variety of sources, including wind, settling ground, earthquakes, gravitational field pull, changing temperatures, and heavy snowfalls.

______________________ 17. The weight of a structure itself and the weight added to the structure under normal use are called ______ loads.

______________________ 18. *True or False?* Dynamic loads include any permanent component of a building, such as the floors, walls, roof, or beams.

19. What is a *live load*?

____

____

____

____

______________________ 20. A horizontal element that spans a gap and rests on two columns is called ______ construction.

21. What is an *arch*?

____

____

____

____

______ 22. Which of the following is an element of an arch that stops the thrust of the arch by tying the ends of the arch to the ground?

A. An abutment
B. A cantilever
C. A post
D. A truss

______________________ 23. *True or False?* Calculating tensile and compressive stress is important for conducting structural analysis.

24. Briefly explain *buckling*.

____

____

____

_______ 25. Which of the following can be expressed as the change of length of an object, divided by the initial length of the object experiencing stress?
A. Buckling
B. Force
C. Strain
D. Young's Modulus of Elasticity

26. What is a *point load*?

____

____

____

_______________________ 27. A force that resists the applied load at the beam supports is called a(n) _______ force.

Match the types of diagrams to the corresponding letters in the figure below.

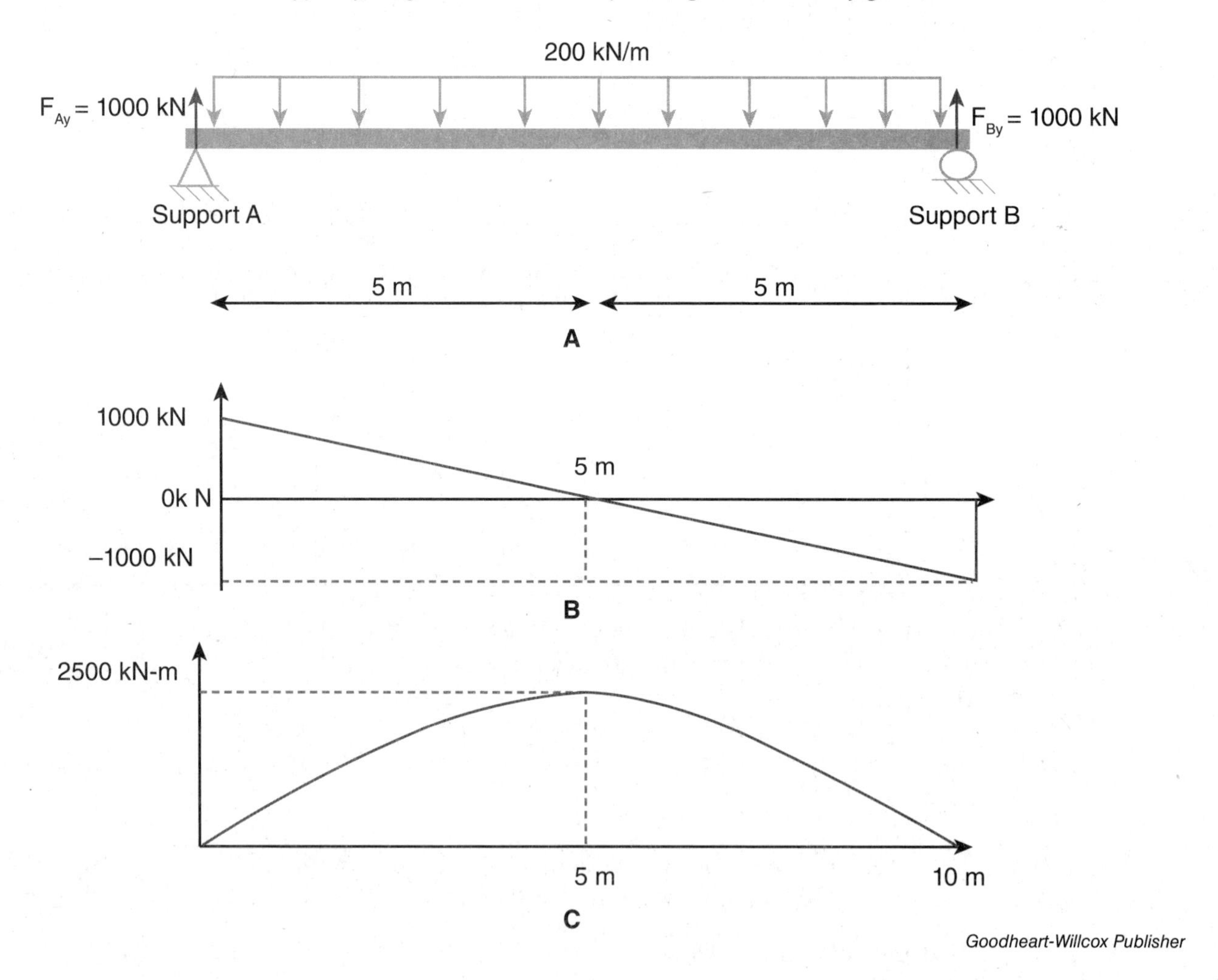

Goodheart-Willcox Publisher

_______ 28. Free body diagram

_______ 29. Moment diagram

_______ 30. Shear force diagram

Name ______________________ Date __________ Class ______________

CHAPTER 19

Harnessing and Using Energy

Complete the following questions and problems after carefully reading the corresponding textbook chapter.

______________ 1. ______ is the ability to do work.

______ 2. The application of a force that moves a mass a distance in the direction of the applied force is called ______.
A. kinetic energy
B. potential energy
C. power
D. work

______ 3. ______ is energy that an object possesses in relation to its motion or due to its position.
A. Chemical energy
B. Mechanical energy
C. Radiant energy
D. Thermal energy

______________ 4. *True or False?* Thermal energy is sometimes called light energy.

5. What is *chemical energy*?

____

____

____

6. What is *nuclear energy*?

____

____

____

C 7. A process in which two atoms are combined into a new, larger atom, releasing large amounts of energy is called ______.

A. biochemical conversion
B. fission
C. fusion
D. radiation

8. Briefly explain what *exhaustible energy resources* are.

They are materials that cannot be replaced

Match the steps of the water cycle to the corresponding letters in the following figure.

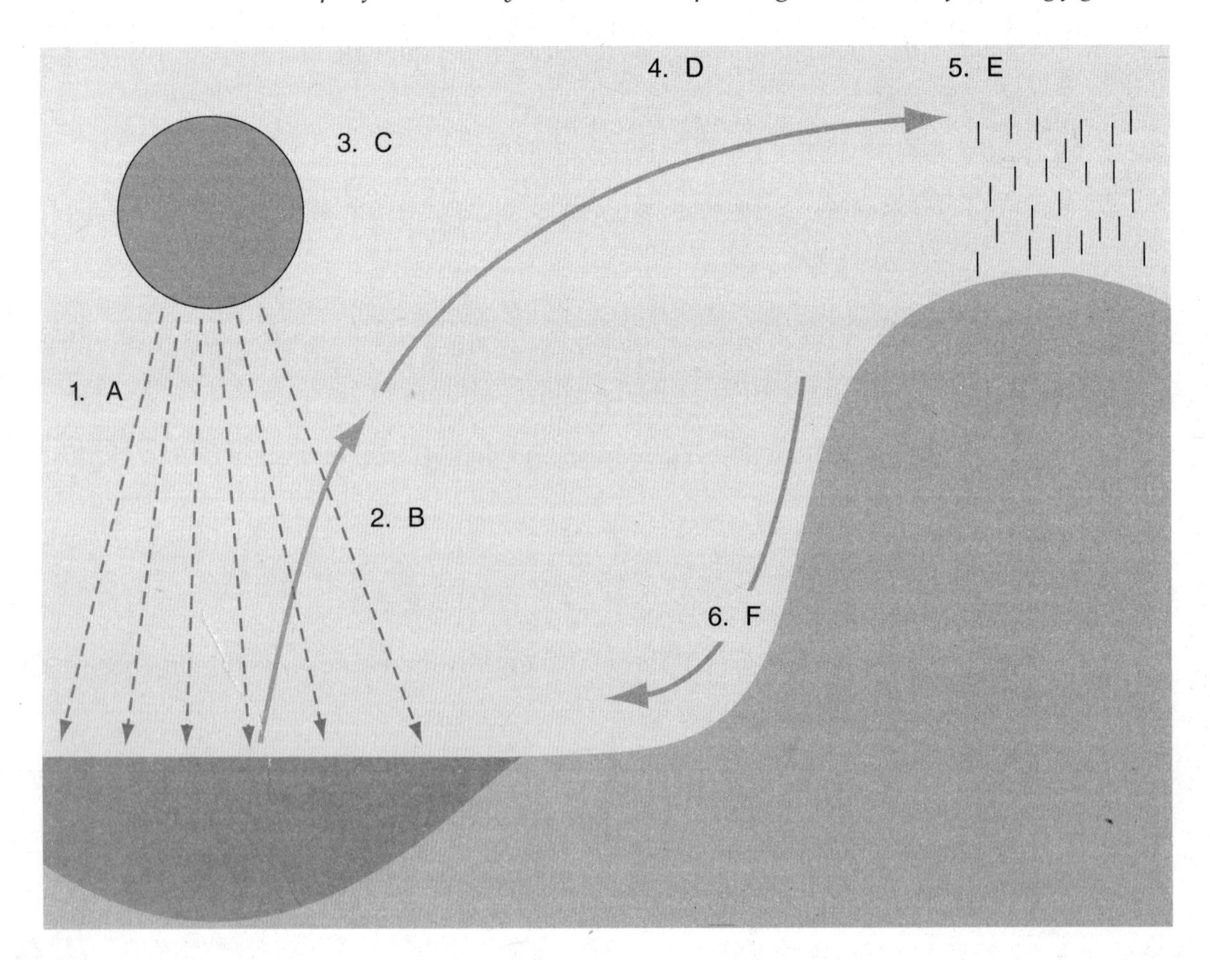

Goodheart-Willcox Publisher

A 9. Radiant energy heats the seawater.

C 10. Vapor forms clouds.

F 11. Water drains off land through rivers to the ocean.

E 12. Water falls as rain or snow.

B 13. Water vapor rises into the atmosphere.

D 14. Wind carries clouds inland.

B 15. Which of the following is *not* a biomass resource?
A. Animal waste
B. Plastic bottles
C. Sewage
D. Straw

biofuels 16. Fuels created from the burning of organic materials are called _______.

True 17. *True or False?* Biogas is gas formed by the breaking down of organic wastes by bacteria in the absence of oxygen.

18. Name three categories of energy-conversion systems.

Renewable-energy converters
Thermal-energy converters
Electrical-energy converters.

C 19. A ______ is a device used to power electric generators in hydroelectric power plants, and consists of a series of blades arranged around a shaft.
A. heat pump
B. solar collector
C. water turbine
D. windmill

insolation 20. The solar energy available in a specific location at any given time is called _______.

True 21. *True or False?* Active collectors directly collect, store, and distribute the heat they convert from solar energy.

22. Briefly explain a *direct-gain solar system.*

This system allows the radiant energy to enter the home through windows, heating inside surfaces

False 23. *True or False?* A photovoltaic cell converts chemical energy into electricity.

24. Briefly describe *geothermal energy.*

It is heat originating in the earth's molten core.

Thermal 25. Ocean _______ -energy conversion systems use the differences in temperature between the various depths of the ocean.

26. What is a *biomass resource*?

____

____

____

______________ 27. Sawdust, bark, logging slash, wood shavings, scrap lumber, and paper are resources from the ______ -products industry.

______ 28. ______ produces a chemical reaction through the application of heat.

A. Anaerobic digestion
B. Biochemical conversion
C. Fermentation
D. Thermochemical conversion

______________ 29. The process in which an organism, such as yeast or bacteria, metabolizes a carbohydrate, such as a starch or sugar, and converts it into an acid or alcohol is called ______.

______________ 30. All ______ engines can be classified as either internal combustion engines or external combustion engines.

Match the strokes of a four stroke cycle-engine to the corresponding letters in the following figure.

Intake valve
Fuel-air mixture
Piston
Crankcase
Compressed mixture
Burning gases
Spark plug
Exhaust valve
Burned gases

A Fuel-air mixture is forced into the cylinder

B Fuel-air mixture is compressed in the cylinder

C Spark plug ignites the fuel-air mixture

D Burned gases are forced out of the cylinder

Goodheart-Willcox Publisher

______ 31. Compression stroke

______ 32. Exhaust stroke

______ 33. Intake stroke

______ 34. Power stroke

______________ 35. *True or False?* Most external combustion engines are steam engines.

______________ 36. The movement of heat along a solid material or between two solid materials touching each other is called ______.

________________ 37. ______ is the transfer of heat between or within fluids, involving the actual movement of the substance.

______ 38. ______ uses electromagnetic waves to transfer heat.
A. Conduction
B. Convection
C. Insolation
D. Radiation

________________ 39. A(n) ______ is the standard device used to capture heat from the atmosphere.

40. State the laws of magnetism and electromagnetism.

____

____

____

____

____

______ 41. The stationary outer magnet in an electric motor is called the ______.
A. armature
B. field magnet
C. north pole
D. south pole

________________ 42. A water-powered plant, called a(n) ______ generating plant, uses a dam to develop a water reservoir.

43. Briefly explain what an armature is in an electric generator.

____

____

____

____

________________ 44. A(n) ______ reduces the voltage of electrical current.

Name ______________________ Date ________ Class ______________

CHAPTER 20

Meeting Needs through Mechanical Engineering

Complete the following questions and problems after carefully reading the corresponding textbook chapter.

1. Define *mechanical engineering*.

____

____

____

____

2. Give two examples of early civilizations that experimented with principles of mechanical engineering.

____

____

______________________ 3. Mechanical engineers need a strong foundation in ______ principles and mathematics.

______ 4. Many of the problems mechanical engineers solve require an understanding of how ______ and motion result in work.
   A. forces
   B. thermodynamics
   C. time
   D. torque

______________________ 5. The greater you kick, pull, or push an object is called an increase in the ______ of the force applied.

______________________ 6. *True or False?* Direction is an element used to describe a force.

7. What is another name for Newton's First Law of Motion?

____

______ 8. Newton's ______ Law of Motion describes the applied force on an object, in relationship to the object's mass and acceleration.
A. First
B. Second
C. Third
D. Fourth

Match the types of motion to the corresponding letters in the following figure.

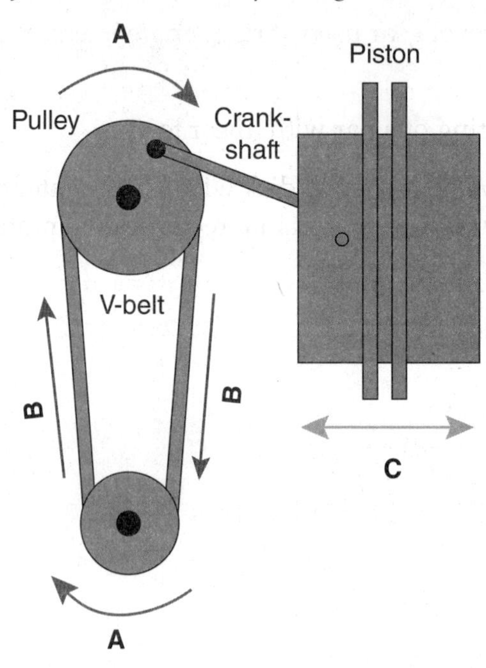

Goodheart-Willcox Publisher

______ 9. Linear motion

______ 10. Reciprocating motion

______ 11. Rotary motion

______________ 12. The length of time required for each cycle of a reciprocating motion to return to its original position is called the ______.

______________ 13. *True or False?* Power transmission takes the energy generated by a converter and changes it into motion.

______________ 14. The analysis of the behavior of solid materials or systems when they are subjected to stresses, loads, and other external forces is called ______.

______ 15. A ______ is an arm attached to a shaft that rotates.
A. cam
B. crank
C. gear
D. lever

16. If a gear system increases the driven force and reduces its speed, is the drive gear being used smaller or larger than the driven gear?

____

Match each term to its correct description.

A. Idler
B. Lever
C. Miter
D. Rack and pinion

_______ 17. A simple machine that changes the direction or intensity of a linear force

_______ 18. A gear type that transfers motion and direction but does not modify speeds

_______ 19. A unique type of bevel gear used at right-angle intersecting shafts to transmit motion and power

_______ 20. A gear type consisting of a bar with linear teeth

_______________ 21. A pear-shaped disk with an off-center pivot point used to change rotating motion into reciprocating motion is called a(n) _______.

_______________ 22. *True or False?* A V-belt consists of a trapezoidal cross section and is comprised of rubber or molded fabric that allows for a bending action.

23. What is fluid mechanics?

____

____

____

_______________ 24. Liquids are used in _______ systems.

_______________ 25. *True or False?* As mechanical engineers solve problems, they often consider and apply principles of pneumatics.

_______________ 26. _______ is the study of heat and temperature and the relation of these factors to work, energy, and power.

_______________ 27. Mechanical engineers can use _______ law to calculate conductive heat transfer.

_______ 28. As mechanical engineers design and test mechanical systems, they measure _______ to ensure their products work effectively and efficiently.

A. power
B. torque
C. work
D. All of the above.

_______________ 29. A measurement that is one newton per meter is called a(n) _______.

Match the uses of hydraulic systems to the corresponding letters in the following figure.

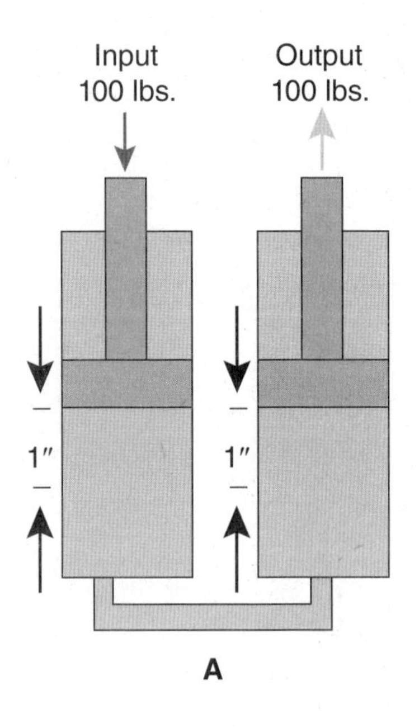

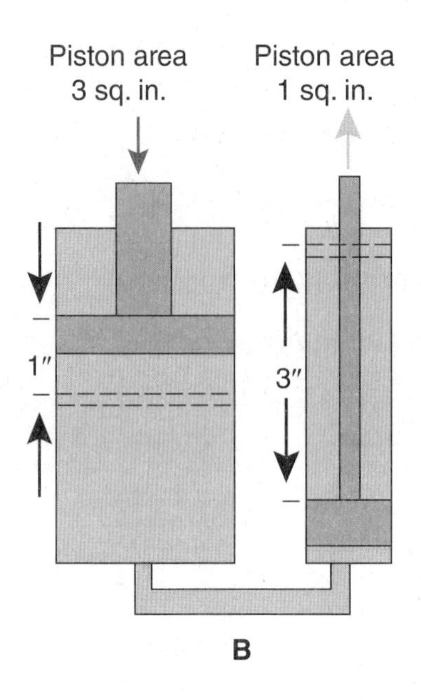

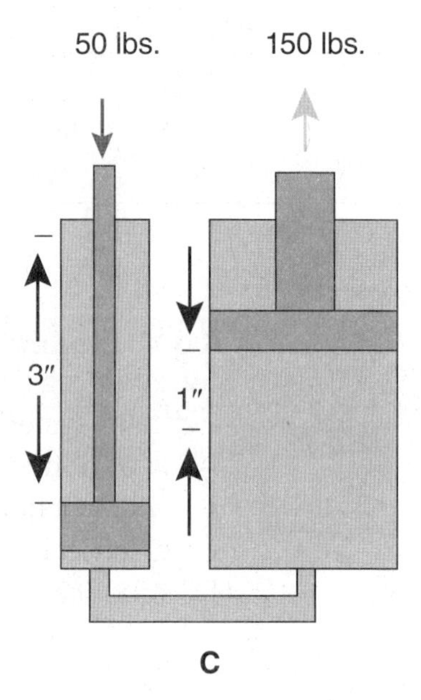

Goodheart-Willcox Publisher

_______ 30. Direction changer

_______ 31. Distance multiplier

_______ 32. Force multiplier

_______________________ 33. *True or False?* Measuring the rate at which work is done gives you the torque.

34. What is *horsepower*?

____

____

____

____

35. What is *torque*?

____

____

____

_______ 36. _______ failure is the failure of an object that occurs when loads or forces have been applied to an object and the object either breaks or is deformed due to some criterion.
A. Dynamic structural
B. Fatigue
C. Static structural
D. Torque

______________________ 37. The failure of an object after being subjected to repeated loading and unloading is ______ failure.

______________________ 38. *True or False?* Mechatronics is a field involving mechanical engineering, electrical engineering, computer control, and information technology.

______________________ 39. Mechanical engineers assist in the ______ industry by applying knowledge of fluid mechanics, designing turbines that transfer heat, and working to create or improve control systems.

______ 40. In the ______ industry, mechanical engineers design heating, ventilation, air conditioning (HVAC) systems; design stress analysis tests; and create elevator and escalator systems.

A. automotive
B. construction
C. electrical power generation
D. robotics

41. In what ways are mechanical engineers employed by electrical power generation industries?

____

____

____

____

____

____

______________________ 42. Mechanical engineers are employed by ______ companies to design mechanical systems and components, such as actuators, motors, and sensors.

Name ______________________ Date __________ Class ______________

CHAPTER 21

Communicating Information and Ideas

Complete the following questions and problems after carefully reading the corresponding textbook chapter.

______________ 1. *True or False?* The purpose of communication is to transmit information.

2. What is *data*?

____

____

____

______________ 3. Advancements in information and ______ technologies have classified our current era as the information age.

4. What is *communication*?

____

____

____

5. Name three basic goals of communication.

____

____

____

______________ 6. *True or False?* Infotainment is information provided in an entertaining way.

7. What is *edutainment*?

____

____

____

Match the parts of the basic communication model to the corresponding letters in the figure below.

Basic Communication Model

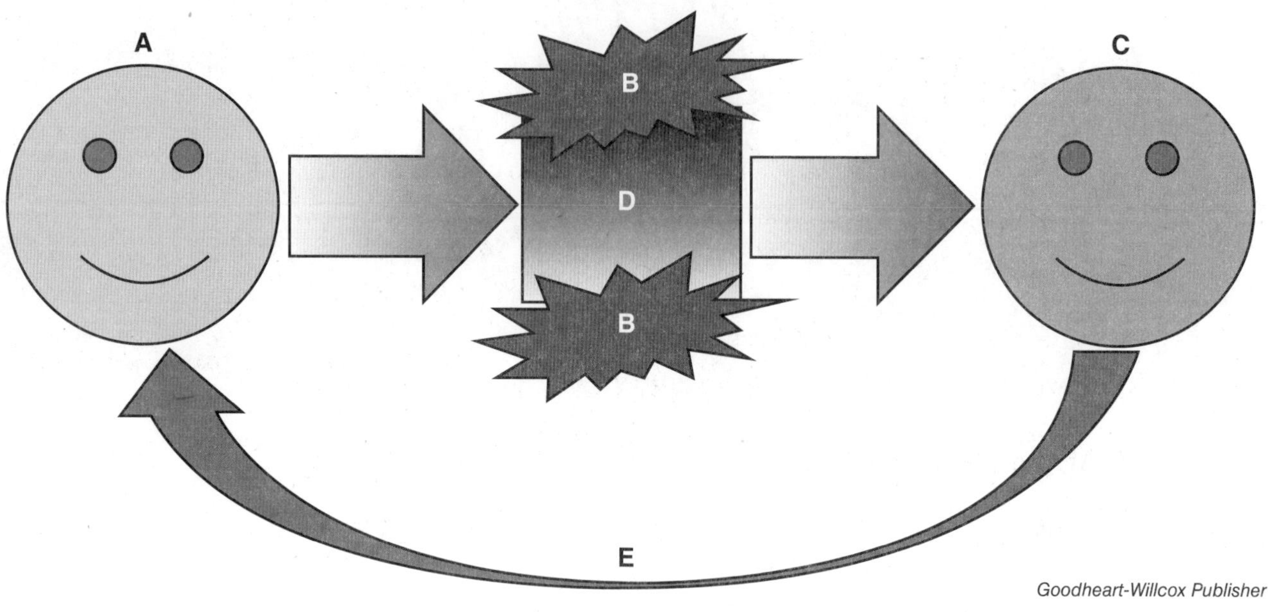

Goodheart-Willcox Publisher

_______ 8. Feedback

_______ 9. Message

_______ 10. Noise

_______ 11. Receiver

_______ 12. Source

13. What is *photographic communication*?

____

____

____

_______ 14. ______ frequencies are used for a wide range of communication systems, including police and fire department radio, cellular telephone, and television communication.
A. Amplitude
B. Broadcast
C. Extremely high
D. Extremely low

_______ 15. A continuous signal that varies in strength or frequency is called a(n) ______ signal.
A. analog
B. broadcast
C. digital
D. frequency

Match the terms to the corresponding letters in the following figure.

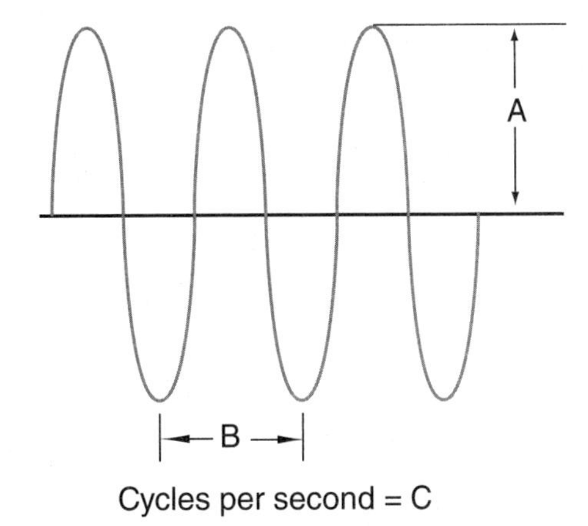

Goodheart-Willcox Publisher

_______ 16. Amplitude

_______ 17. Cycle

_______ 18. Frequency

_______________ 19. *True or False?* The two main types of telecommunication systems are hardwired systems and broadcast systems.

_______________ 20. A radio broadcast system that encodes the message on the carrier wave by changing the wave's frequency is called frequency _______.

_______ 21. _______ is a set of instructions for a computer and its applications.

A. Hardware
B. A network
C. A server
D. Software

_______________ 22. *True or False?* A server is a group of computing devices that are connected together and share resources.

_______ 23. A line, typically fiber-optic, that connects regions in a company's information technology (IT) systems is called a _______.

A. backbone
B. network access point (NAP)
C. point of presence (POP)
D. router

24. What is a *router*?

____

____

____

Match the steps for accessing information to the corresponding letters in the following figure.

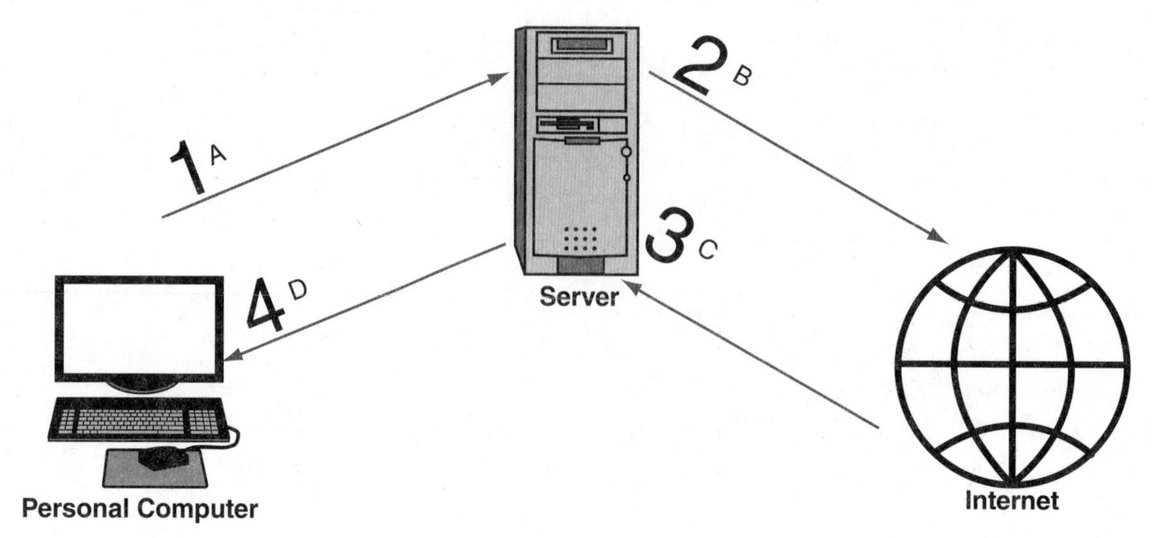

Goodheart-Willcox Publisher

_______ 25. Internet service provider (ISP) host computer connects with the Internet.

_______ 26. ISP host computer requests and receives information.

_______ 27. ISP host computer sends information.

_______ 28. User connects with ISP host computer.

_______ 29. An identifying number assigned to each computer connected to the Internet is called a(n) _______.
A. Internet access number
B. Internet domain
C. Internet Protocol (IP) address
D. ISP

30. Name four of the original types of websites.

____

____

_______________________ 31. The protocol that allows a user to retrieve and modify files on another computer connected to the Internet is the file-_______ protocol.

32. Name three common uses of the Internet.

____

____

____

Name ______________________________

______ 33. A(n) ______ is an underlined phrase, button, or other means that can be selected.
- A. hyperlink
- B. HTML tag
- C. search engine
- D. URL

______________________ 34. A special Internet site that operates on the principle of key words to allow individuals to search the Web by topic is called a search ______.

______________________ 35. *True or False?* A simple mail transfer protocol (SMTP) e-mail server handles incoming mail.

36. Briefly explain the *Internet Message Access Protocol (IMAP).*

____

____

____

____

______________________ 37. ______ involves selling products and services over the Internet.

Name ______________________ Date __________ Class ______________

CHAPTER 22

Meeting Needs through Electrical, Computer, and Software Engineering

Complete the following questions and problems after carefully reading the corresponding textbook chapter.

______ 1. ______ design, test, and oversee the production of devices that use electricity to control, process, store, and receive information.

A. Computer engineers
B. Computer scientists
C. Electrical engineers
D. Software engineers

______________________ 2. Computer engineering involves the development, improvement, and ______ of the production and maintenance of computer hardware and software.

3. What is *software engineering*?

____

____

____

____

______________________ 4. In order to develop and create new electrical devices, computer systems, and computer software, engineers must understand the science of ______.

______________________ 5. The collection of scientific knowledge used to describe the flow of electrical energy through an electrical circuit is called circuit ______.

______________________ 6. *True or False?* Electromagnetic induction is the flow of electric charge.

Match the parts of an atom to the corresponding letters in the following figure.

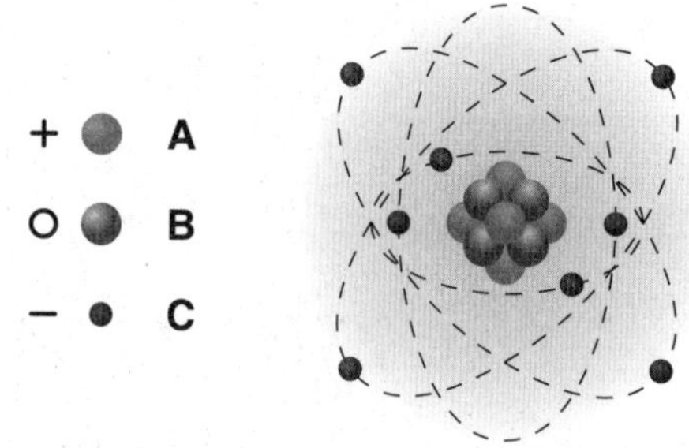

chromatos/Shutterstock.com

_______ 7. Electron

_______ 8. Neutron

_______ 9. Proton

_______ 10. A source of energy in electrical circuits that converts chemical energy to electrical energy is called a(n) _______.
A. battery
B. electrical switch
C. electrolyte
D. light-emitting diode (LED)

_______________ 11. *True or False?* The anode is the positively charged side of the battery.

12. What is an *electrolyte?*

____

____

____

____

_______________ 13. The rate of electron flow through a circuit is the _______.

_______________ 14. *True or False?* Ohm's law defines the relationship among voltage, current, and resistance in an electrical circuit.

_______ 15. The source of the electron flow in an electrical circuit is the _______ source.
A. current
B. electrical
C. power
D. voltage

_______________ 16. An electrical component that interrupts the flow of electrons from a power source in an electrical circuit is an electrical _______.

_______________ 17. *True or False?* A resistor expands the flow of electrons in a circuit and increases the current and voltage.

18. What is a *capacitor*?

____

____

____

_______________________ 19. A special diode that releases energy in the form of photons when a suitable voltage is applied is called a(n) ______ diode.

20. What is an *integrated circuit*?

____

____

____

_______________________ 21. *True or False?* In a parallel circuit, components are connected in a way that provides only a single pathway for electrons to flow.

22. What is *Kirchhoff's voltage law*?

____

____

____

______ 23. A(n) ______ is a circuit in which components are connected across common ends or nodes, providing multiple pathways for electrons to flow.
- A. IC
- B. parallel circuit
- C. series circuit
- D. series-parallel circuit

_______________________ 24. Kirchhoff's ______ law states that the sum of the currents across each branch or loop of a parallel circuit is equal to the total current in a circuit.

_______________________ 25. *True or False?* A series-parallel circuit has components that are arranged in series and components that are arranged in parallel.

26. What is *alternating current (ac)*?

____

____

____

_______________________ 27. The production of electromotive force or voltage by continuously moving a conductor, such as copper wire, through varying magnetic fields is electromagnetic ______.

_______________________ 28. *True or False?* Analog signals are difficult to work with in electronics because of the infinite variations.

_______ 29. Which of the following enables complex decisions to be made within computer systems based on a series of yes or no questions?

A. An algorithm
B. Binary code
C. Digital logic
D. A programming language

_______ 30. A(n) _______ gate produces an output signal of 1, if both input signals are 1.

A. AND
B. OR
C. NAND
D. NOT

_______________ 31. A(n) _______ gate produces an output signal of 1, if at least one of the input signals is 1.

_______________ 32. *True or False?* An XNOR gate produces an output signal of 1, if only one input signal is 1.

33. What is *sequential logic*?

____

____

____

_______________ 34. The _______ is the working part of the computer that carries out instructions.

_______________ 35. Both the computer and the user can read or change _______ memory.

_______ 36. Instructions that direct a computer to perform specific tasks are called _______.

A. software
B. hardware
C. memory
D. operating systems

_______________ 37. A language that consists of only 1s and 0s is _______.

_______________ 38. *True or False?* In binary code, *0* means *on*.

_______ 39. In binary code, a _______ means *off.*

A. 0
B. 1
C. 2
D. blank space

_______________ 40. A computer component that allows the CPU to access different information or data within the computer's memory by storing the location of the information is the _______.

_______________ 41. A(n) _______ is a set of step-by-step instructions used to perform calculations, process data, and automatically answer questions.

42. Name six popular programming languages.

____

____

____

Name ______________________________ Date __________ Class ____________________

CHAPTER 23

Transporting People and Cargo

Complete the following questions and problems after carefully reading the corresponding textbook chapter.

1. What is *transportation*?

____

____

____

____________________ 2. ______ systems have been developed for land, water, air, and space.

3. List the three major types of land-transportation systems.

____

____

____

____________________ 4. *True or False?* Continuous-flow systems include freight, passenger, and mass-transit systems.

5. What is a *space-transportation system*?

____

____

____

______ 6. Which of the following is *not* a major component in each transportation system?
A. Pathways
B. Support structures
C. Utilities
D. Vehicles

__________ 7. Locations where transportation activities begin and end are ______.

8. Name the five systems common to all vehicles.

____

____

____

__________ 9. *True or False?* The suspension system produces proper support for the weight of a vehicle and its cargo as the vehicle moves along its pathway.

__________ 10. Speed and direction are the two types of ______ systems.

__________ 11. ______ transportation includes all movement of people and goods on or under the surface of the earth.

______ 12. Most ______ vehicles move along their pathways on rolling wheels.
A. air-transportation
B. land-transportation
C. space-transportation
D. water-transportation

Match the engine strokes to the corresponding letters in the following figure.

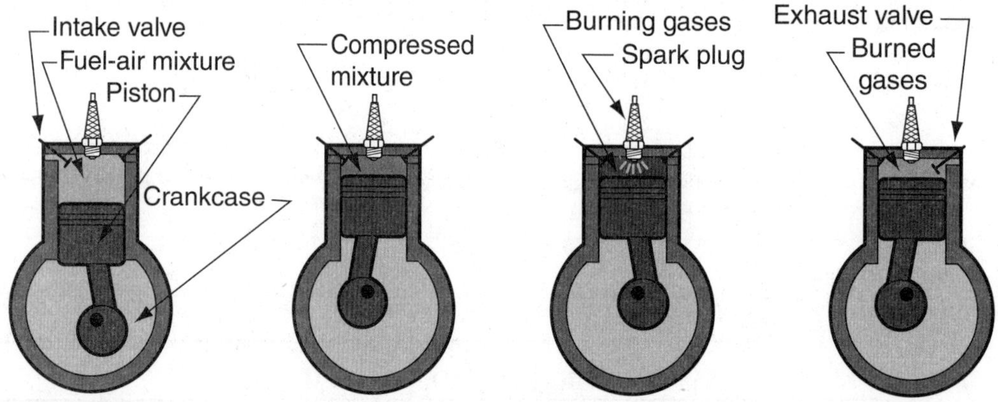

A Fuel-air mixture is forced into the cylinder

B Fuel-air mixture is compressed in the cylinder

C Spark plug ignites the fuel-air mixture

D Burned gases are forced out of the cylinder

Goodheart-Willcox Publisher

______ 13. Compression stroke

______ 14. Exhaust stroke

______ 15. Intake stroke

______ 16. Power stroke

17. List the three major parts of a gasoline-electric hybrid propulsion system.

____

____

Match the components of a power-transmission system to the corresponding letters in the following figure.

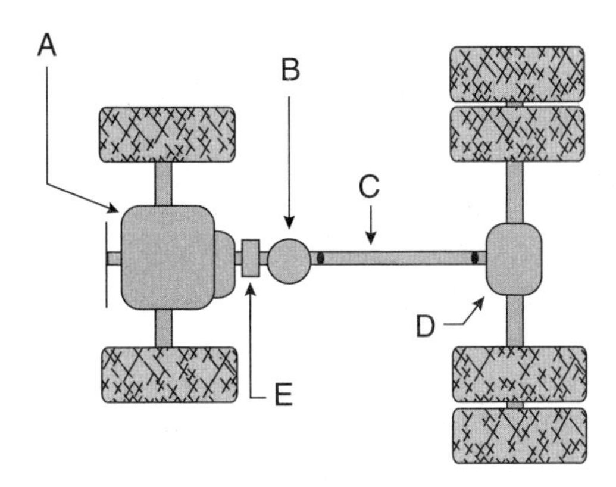

Goodheart-Willcox Publisher

_______ 18. Clutch

_______ 19. Differential

_______ 20. Drive shaft

_______ 21. Engine

_______ 22. Transmission

_______________ 23. *True or False?* A manual transmission is a transmission system that uses valves to change hydraulic pressure so the transmission shifts its input-to-output ratios.

24. What is a *transaxle*?

____

____

____

_______________ 25. _______ carry the load of the vehicle to the wheels.

_______________ 26. Water transportation on oceans and large inland lakes is called _______ shipping.

27. What is a *merchant ship*?

____

____

____

_______ 28. A(n) _______ is used to power small boats and attaches to the stern.
A. steam turbine
B. outboard motor
C. hydrofoil
D. sail

_______________________ 29. *True or False?* Buoyancy is the upward force exerted on an object immersed in a fluid.

30. What is a *hovercraft*?

____

____

____

_______________________ 31. A(n) _______ is a prop mounted at a right angle to the keel.

32. What is a *lighter-than-air vehicle*?

____

____

____

_______ 33. Most passenger and cargo aircraft are _______.
A. blimps
B. fixed-wing aircraft
C. lighter-than-air vehicles
D. rotary-wing aircraft

_______________________ 34. The body of an aircraft that carries people or cargo is called the _______.

_______________________ 35. *True or False?* The type of engine that powers the majority of business and commercial aircraft is the internal combustion engine.

36. Name the four major forces affecting an airplane's ability to fly.

____

____

____

____

_______________________ 37. *True or False?* An airfoil separates the air into two streams to create lift.

_______________________ 38. The angle of the blades on a helicopter can be _______ to generate additional lift.

Identify the components of a turbojet engine in the following figure.

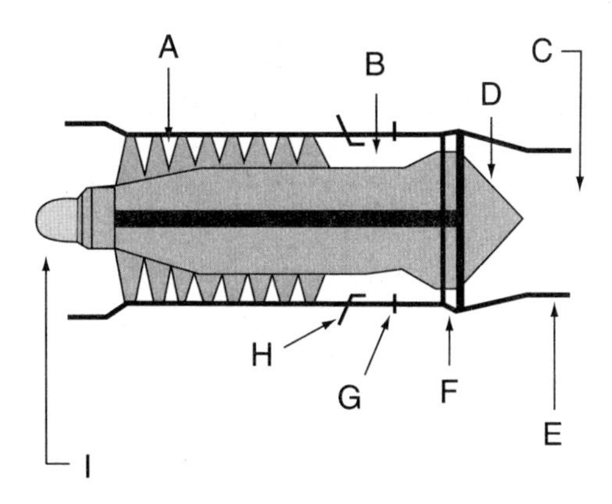

Goodheart-Willcox Publisher

_______ 39. Jet nozzle

_______ 40. Compressor

_______ 41. Fuel nozzle

_______ 42. Turbine

_______ 43. Combustion chamber

_______ 44. Tail cone

_______ 45. Igniter

_______ 46. Which of the following is *not* a function or system of avionics?
A. Maintenance systems
B. Automatic pilot
C. Communication systems
D. An instrument landing system (ILS)

______________________ 47. Space travel can be classified as unmanned or manned flight and _______ or outer space travel.

48. List the three major parts of a space vehicle and its launch systems.

____

____

____

____

49. What is a *solid-fuel rocket*?

____

____

____

____

Match the layers of the earth's atmosphere to the corresponding letters in the figure below.

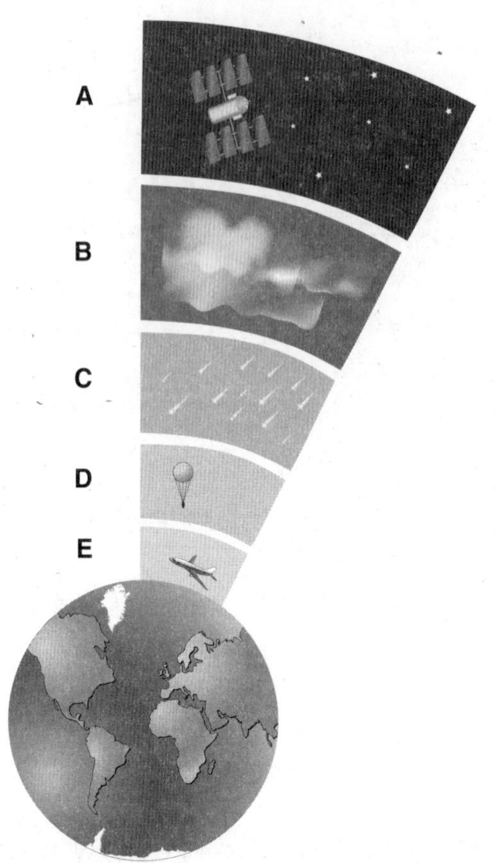

Designua/Shutterstock.com

_______ 50. Exosphere

_______ 51. Mesosphere

_______ 52. Stratosphere

_______ 53. Thermosphere

_______ 54. Troposphere

Name ______________________________ Date __________ Class ____________________

CHAPTER 24

Meeting Needs through Aerospace Engineering

Complete the following questions and problems after carefully reading the corresponding textbook chapter.

______ 1. ______ engineering applies mathematical, scientific, and engineering knowledge to the development of aircraft and spacecraft and their related technologies.
- A. Aeronautical
- B. Aerospace
- C. Astronautical
- D. None of the above.

______________________ 2. *True or False?* Many centuries of scientific and mathematical breakthroughs and developments led to the creation of aerospace engineering.

3. What is *biomimicry*?

____

____

____

4. An ornithopter is a flying machine designed by ______.

______________________ 5. *True or False? Sputnik II* carried a living passenger.

6. What is *aeronautical engineering*?

____

____

____

____

Match the forces to the corresponding letters in the following figure.

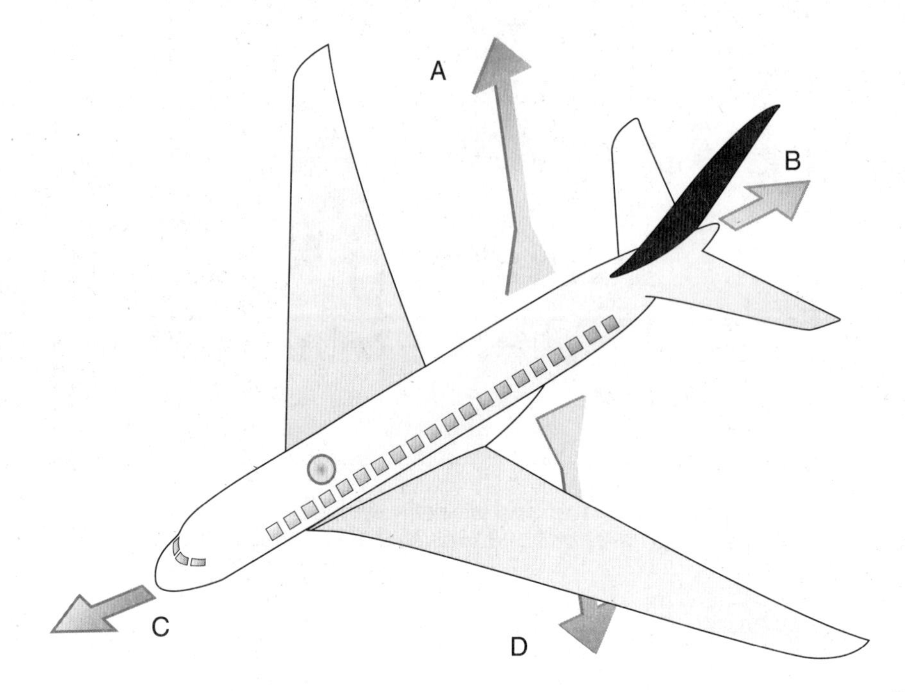

Goodheart-Willcox Publisher

_______ 7. Drag

_______ 8. Lift

_______ 9. Thrust

_______ 10. Weight

_______ 11. The characteristic curvature along the upper and lower surfaces of an airfoil is known as _______.

A. aspect ratio
B. camber
C. chord
D. span

span _______ 12. The measurement from wing tip to wing tip is called _______.

13. What is *aspect ratio*?

The relationship b/t the span and the chord

incidence _______ 14. The angle at which the wings are attached to the fuselage of the plane is the angle of _______.

True _______ 15. *True or False?* There is no one-size-fits-all airfoil design.

______ 16. Spinning ______ provide the lift and thrust needed for a helicopter to leave the ground.
A. airfoils
B. cambers
C. chords
D. rotors

______________________ 17. The ______ is comprised largely of air and surrounds Earth.

18. What is *fluid dynamics*?

____

____

____

______ 19. ______ fluid often sticks to a moving object's surface.
A. Aerodynamic
B. Frictional
C. Laminar
D. Viscous

______ 20. Which of the following states that an increase in fluid speed creates a decrease in pressure?
A. Bernoulli's principle
B. The law of conservation of energy
C. Newton's law of motion
D. The Venturi effect

______________________ 21. A law that states that energy cannot be created or destroyed is the ______ of energy.

______________________ 22. *True or False?* Sir Isaac Newton's third law of motion states that an object will remain at rest until a force is applied to it.

______________________ 23. To increase efficiency, engineers choose structural materials that can reduce ______.

Name ______________________________ Date __________ Class ____________________

CHAPTER 25

Medical and Health Technologies

Complete the following questions and problems after carefully reading the corresponding textbook chapter.

____________________ 1. ______ involves actions that keep the body healthy.

____________________ 2. *True or False?* Aerobic exercise involves heavy work by a limited number of muscles.

______ 3. Which of the following uses oxygen to keep large muscle groups moving continuously?
A. Aerobic exercise
B. Anaerobic exercise
C. Interventional exercise
D. Therapeutic exercise

____________________ 4. Playing fields and venues for sports are part of our built environment and a result of ______ technology.

5. What is *chronic traumatic encephalopathy*?

____

____

____

____

______ 6. Any change interfering with the appearance, structure, or function of the body is a(n) ______.
A. disease
B. immunization
C. intervention
D. medicine

7. List five health-care professions.

____

____

____

8. What is *immunization*?

____

____

____

____

_______________________ 9. ______ technologies are health-care devices, products, processes, and systems used to diagnose, monitor, or treat medical conditions.

______ 10. Devices that measure oxygen in the blood, listen to heart rhythm, and measure blood pressure are examples of ______ diagnostic equipment.
A. invasive
B. minimally invasive
C. noninvasive
D. routine

_______________________ 11. The use of electromagnetic waves and ultrasonics to diagnose diseases and injuries is called ______.

Identify the components of a CT scanner in the following figure.

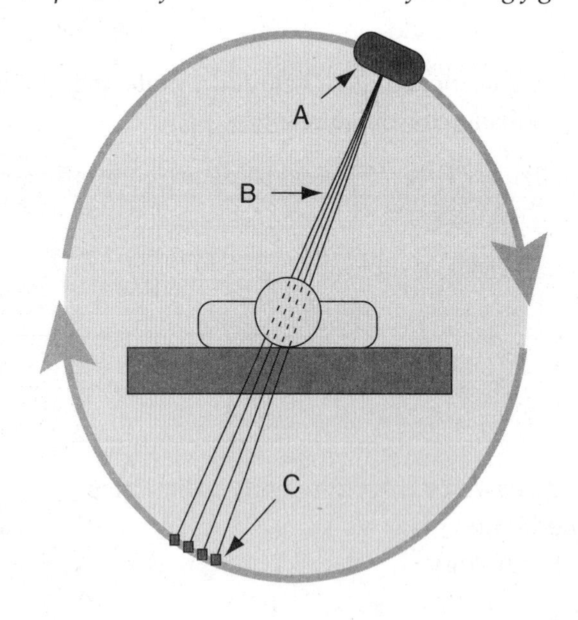

Goodheart-Willcox Publisher

______ 12. X-ray detectors

______ 13. X-ray source

______ 14. X-rays

______ 15. A(n) ______ is a visual records of the heart's electrical activity.
A. computerized tomography (CT)
B. magnetic resonance imaging (MRI)
C. electrocardiograph
D. ultrasound

______________ 16. *True or False?* Invasive diagnostic equipment is used to take a tissue sample for laboratory examination

17. What is an *endoscope*?

____

____

____

______ 18. Which of the following is *not* a way in which drugs are typically classified?
A. Illness or condition treated
B. Method of administering
C. Method of dispensing
D. Trial phase

19. What is a *prescription drug*?

____

____

____

______________ 20. A research technique in which neither the group being tested, nor the researchers, know who is receiving the active substance and who is receiving the placebo is called a(n) ______.

______________ 21. Synthetic drugs developed by altering the structure of existing substances are called ______.

______________ 22. *True or False?* Clinical trials are generally conducted in three phases.

23. What is *interventional radiology*?

____

____

____

______ 24. The treatment of diseases or disorders with radiation is ______.
A. emergency medicine
B. interventional radiology
C. surgery
D. therapeutic radiology

______________________ 25. ______ engineering applies mechanical engineering principles and materials to surgery and prosthetics.

26. What is *telemedicine*?

____

____

____

Name ______________________ Date __________ Class ______________

CHAPTER 26

Meeting Needs through Biomedical Engineering

Complete the following questions and problems after carefully reading the corresponding textbook chapter.

1. What is *biomedical engineering*?

____

____

____

____

______________ 2. *True or False?* Biomedical engineering activities began in very recent years, and this engineering discipline is expected to grow slowly over time.

______ 3. A biomedical engineer ______.
A. devises new and better medical technologies
B. designs artificial organs
C. works in genetic engineering
D. All of the above.

______________ 4. Tissues or devices that are surgically placed inside or on the exterior of the body are ______.

______________ 5. *True or False?* Many implants are prosthetic devices designed to replace missing or damaged body parts.

6. What is a *transtibial prosthesis*?

____

____

____

_______________ 7. A prosthesis that uses advanced biosensors, as well as mechanical and electrical components, is called a(n) ______ prosthesis.

_______________ 8. Bioartificial organs are produced using ______ tissues.

______ 9. A subdiscipline of biology that studies the structure and function of the human body is ______.

A. biomechanics
B. human physiology
C. kinematics
D. kinesiology

_______________ 10. The cardiovascular system can also be referred to as the ______ system.

_______________ 11. *True or False?* The respiratory system is the system of the body in which the heart, blood, and blood vessels work together to carry oxygen, nutrients, hormones, and waste through the body.

12. What are the four main components of blood?

____

____

_______________ 13. The blood cells that work to fight off infections are ______ blood cells.

_______________ 14. The circuit that carries blood from the heart to the cells of the body is the ______ circuit.

Match the following terms with the corresponding description.

A. Arteries
B. Capillaries
C. Veins

______ 15. Enable the exchange of nutrients, oxygen, and chemicals between the blood and cells of the body

______ 16. Carry blood back to the heart

______ 17. Carry blood away from the heart

______ 18. The ______ system brings oxygen into the body, delivers it to cells, and rids the body of carbon dioxide.

A. cardiovascular
B. circulatory
C. musculoskeletal
D. respiratory

_______________ 19. The ______ is the breathing muscle that contracts and expands the airtight cavity surrounding the lungs.

_______ 20. The _______ are two air tubes that branch off of the trachea into each lung.
A. bronchi
B. diaphragms
C. epiglottises
D. pharynxes

_______________ 21. *True or False?* Diffusion is the movement of molecules from a region of high concentration to a region of low concentration.

22. What is the *musculoskeletal system*?

____

____

____

_______________ 23. The section of the skeleton that provides support and protection for the brain, spinal cord, and chest cavity is the _______ skeleton.

_______________ 24. *True or False?* The appendicular skeleton is comprised of 126 bones.

_______ 25. Which of the following are fixed joints that do not move?
A. Condylar joints
B. Ossicles
C. Saddle joints
D. Sutures

_______________ 26. An elastic, fibrous tissue that can contract to produce movement is called a(n) _______.

27. What is *biomechanics*?

____

____

____

_______ 28. The branch of mechanics that deals with the analysis of objects that are accelerating as a result of acting forces is called _______.
A. dynamics
B. kinematics
C. kinetics
D. statics

_______________ 29. The compatibility between the properties of materials used in biomechanical devices and the living systems in which they are used is called _______.

30. What is a *biomaterial*?

____

____

____

False 31. *True or False?* Tissue engineering is the manipulation of the genetic material of living organisms.

B 32. Which of the following is *not* an example of biomedical imaging technology?
A. Radiography
B. Medical implants
C. Magnetic resonance imaging
D. Ultrasound

government 33. ______ agencies are responsible for regulating the practices involved with biomedical engineering.

34. List the two primary agencies in the United States that create guidelines and standards for testing and implementing new biological, medical, and health technologies.

-The Food and Drug Administration (FDA)
-The National Institutes of Health (NIH)

35. What is *medical ethics*?

Ethics surrounding the issues of bio[illegible] engineering affecting the well being of others.

Name ________________________________ Date ___________ Class ____________________

CHAPTER 27

Agricultural and Related Biotechnologies

Complete the following questions and problems after carefully reading the corresponding textbook chapter.

______ 1. ______ is people using materials, information, and machines to produce food and natural fibers.

A. Agriculture
B. Animal husbandry
C. Aquaculture
D. Crop production

____________________ 2. *True or False?* Harvesting is the process of growing plants for various uses, including food for humans, feed for animals, and natural fibers for a variety of uses.

3. What is *animal husbandry*?

____

____

____

______ 4. Which of the following are crops with edible leaves, stems, roots, and seeds that provide important vitamins and minerals for the daily diet?

A. Forage crops
B. Grain
C. Nuts
D. Vegetables

____________________ 5. A plant grown for animal feed is called a(n) ______ crop.

6. Name four major groups of farm equipment used in crop production.

____

____

____

____________________ 7. Wheel tractors with rear power wheels usually have ______ front wheels.

______ 8. A ______ is a piece of tilling equipment that breaks, raises, and turns the soil.
A. combine
B. grain drill
C. plow
D. tractor

____________________ 9. *True or False?* A plowshare is the part of a plow that actually cuts trenches.

____________________ 10. A machine used to control pests and weeds that uses a series of hoe-shaped blades pulled through the ground is called a(n) ______.

11. What are three things an irrigation system must have?

____

____

____

____

____

______ 12. A ______ irrigation system sends large volumes of water across a field.
A. drip
B. flood
C. furrow
D. sprinkler

____________________ 13. A system in which many sprinkler lines are used to cover an entire field to irrigate it all at once is called a(n) ______ sprinkler system.

14. Briefly explain a drip-irrigation system.

____

____

____

______ 15. Bands of hay are called ______.
A. bales
B. furrows
C. swathers
D. windrows

____________________ 16. The most common type of grain elevator is a(n) ______ elevator.

17. What are the two components that all hydroponic systems contain?

____

____

____

______ 18. Which of the following is *not* a technology involved with raising livestock?
A. Animal-waste facilities
B. Buildings and machines used to feed animals
C. Fences and fencing to establish feedlots and pastures
D. Tillage equipment

______________ 19. The height of the fence depends on the size and ______ of the contained animals.

______________ 20. *True or False?* A major challenge for large-scale livestock production is animal-waste disposal.

______________ 21. Traditional fishing is also called ______ fishing.

22. What is *biotechnology*?

____

____

____

______ 23. The process of producing new pest-resistant and chemical-tolerant crops that help combat diseases is ______.
A. agricultural technology
B. gene splicing
C. chemical processing
D. genetic engineering

24. Name the seven basic nutrients provided by food.

____

____

____

____

______ 25. Which of the following makes up about 60% of the human body?
A. Fat
B. Muscle
C. Protein
D. Water

______________ 26. Using knowledge, machines, and techniques to convert agricultural products into foods that have specific textures, appearances, and nutritional properties is called food-______ technology.

27. What is *primary food processing*?

____

____

____

______________________ 28. ______ food processes include both material conversion and food preservation.

______________________ 29. *True or False?* Heat processing is a food processing method that uses machines to change the physical form of a food product.

______ 30. Which of the following is *not* a major step in the milling process?
A. Conditioning
B. Grain receiving and cleaning
C. Roasting
D. Storage and packaging

______________________ 31. The second step in roasting coffee beans is called ______.

______________________ 32. In cheddar cheese manufacturing, the liquid part in the vat with the curd is called ______.

______________________ 33. *True or False?* Some preservation techniques can reduce the quality of food.

34. Name the three commonly used drying methods.

____

____

______ 35. Which of the following preservation methods adds flavor to meat and fish while preserving them?
A. Aseptic packaging
B. Canning
C. Drying
D. Smoking

______________________ 36. Controlled atmospheres can extend the ______ lives of fruits and vegetables by several months.

37. What is *irradiation*?

____

____

____

Name ______________________________

_______ 38. Secondary food-processing technology involves _______ and processing ingredients and food products to change their properties.
A. harvesting
B. separating
C. combining
D. planting

Match the steps in food-product development to the corresponding letters in the following figure.

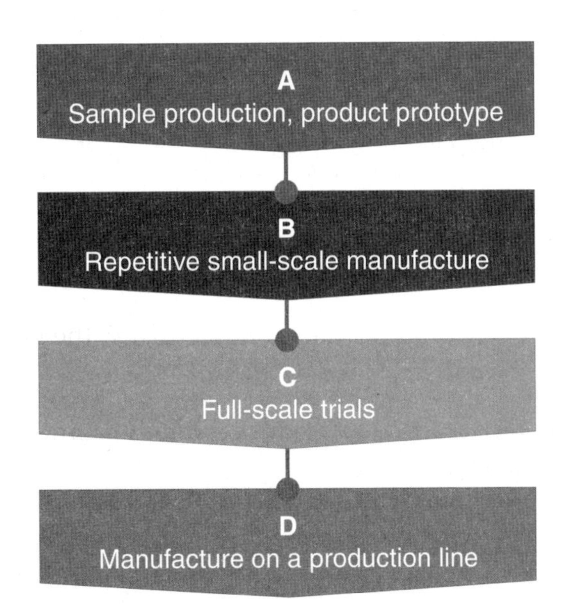

Goodheart-Willcox Publisher

_______ 39. Continuous production

_______ 40. Pilot plant investigation

_______ 41. Pilot production

_______ 42. Small-scale investigation

43. List four of the processes most foods go through during food-product manufacture.

____

____

____

____

_______ 44. Which of the following is *not* included in pasta production?
A. Blending the ingredients
B. Drying the product
C. Kneading and mixing the dough
D. Roasting

Name ______________________________ Date __________ Class ______________

CHAPTER 28

Meeting Needs through Chemical Engineering

Complete the following questions and problems after carefully reading the corresponding textbook chapter.

1. What is *chemical engineering*?

____

____

____

____

__________________ 2. *True or False?* Fermentation is a process in which an organism metabolizes an acid or alcohol and converts it into a carbohydrate.

__________________ 3. Deoxyribonucleic acid (DNA) is the ______ material of living organisms.

______ 4. Which of the following is a chemical made from crude oil?
A. A biochemical
B. A halogen
C. An organic chemical
D. A petrochemical

__________________ 5. Professionals who develop new materials and chemicals by using energy to create a basic chemical change in raw materials or chemicals are called ______.

6. What is *physical science*?

____

____

____

_______ 7. Any substance that has mass or takes up space is _______.
A. an atom
B. a chemical element
C. energy
D. matter

8. What is an *atom*?

____

____

____

_______ 9. Which of the following has a positive electric charge?
A. An anion
B. An electron
C. A neutron
D. A proton

10. What is an *electron orbital*?

____

____

____

____

_______________________ 11. *True or False?* The Aufbau principle determines how the electrons are configured in each orbital.

_______________________ 12. An electron in the outermost shell of an atom is called a(n) _______ electron.

Match the terms to the corresponding description or definition.

A. Atomic number
B. Chemical name
C. Chemical symbol
D. Group
E. Period
F. Chemical elements

_______ 13. Column on the periodic table of elements

_______ 14. Building blocks of matter

_______ 15. Indicates the number of protons in the nucleus of element's atom

_______ 16. Row on the periodic table of elements

_______ 17. Letter(s) representing an element

_______ 18. Which of the following is a negatively charged ion?

A. An allotrope
B. An anion
C. A cation
D. An electron

______________ 19. All elements within the same period have the same number of _______.

_______ 20. Any substance consisting of two or more different atoms that are bonded or stuck together is called a(n) _______.

A. allotrope
B. chemical compound
C. chemical element
D. molecule

21. What is an *alkali metal*?

____

____

____

_______ 22. Which of the following are the chemical elements that have full valence orbitals, are stable, and typically do not react with other elements?

A. Alkali metals
B. Alkaline earth metals
C. Halogens
D. Noble gases

______________ 23. A chemical element that has seven electrons in its valence orbital and readily gains electrons and reacts with metals to form chemical compounds is called a(n) _______.

______________ 24. The _______ dictates the element's properties in a chemical reaction.

25. What is a *covalent bond*?

____

____

____

_______ 26. _______ is the chemical reaction that causes steel to rust.

A. Cation
B. Fermentation
C. Oxidation
D. Transition

______________ 27. A standard method to represent what occurs in a chemical reaction is called a chemical _______.

28. In a chemical equation, what is a *product*?

____

____

____

_______ 29. Which of the following is a part of chemistry that studies the amount of matter involved in a chemical reaction?
A. The Aufbau principle
B. Biochemistry
C. Organic chemistry
D. Stoichiometry

Match the states of matter to the corresponding letters in the figure below.

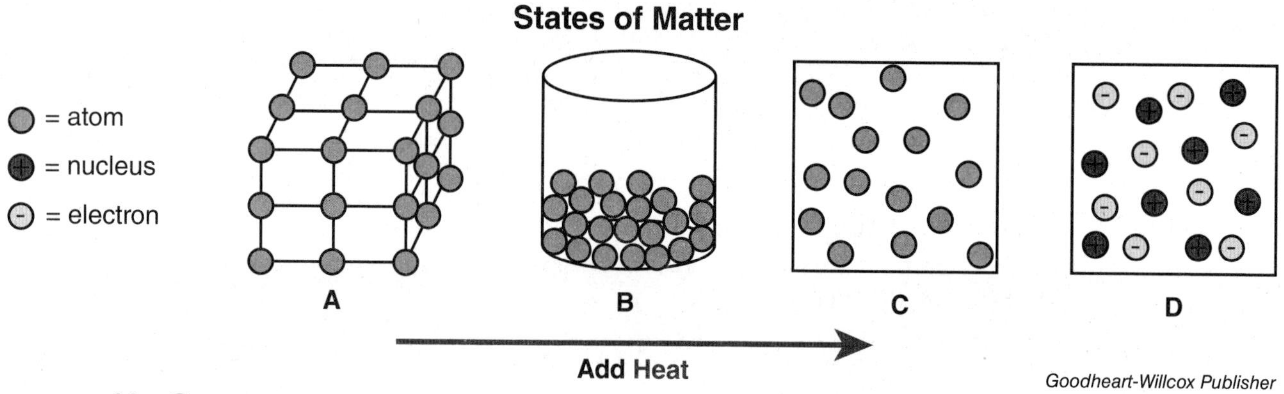

States of Matter

= atom
= nucleus
= electron

A B C D

Add Heat

Goodheart-Willcox Publisher

_______ 30. Gas

_______ 31. Liquid

_______ 32. Plasma

_______ 33. Solid

_______________ 34. The state of matter that has no definite shape or volume, but that can form structures when subjected to a magnetic field, is _______.

_______________ 35. *True or False?* Allotropes are different forms of the same element in the same physical state.

36. What is *biochemical engineering*?

____

____

____

____

_______________ 37. _______ chemistry is the study of the chemical compounds and processes that take place in living organisms.

Name ______________________________ Date __________ Class ________________

CHAPTER 29

Technology and Engineering: A Societal View

Complete the following questions and problems after carefully reading the corresponding textbook chapter.

______________________ 1. ______ is designed through engineering practices to help people modify and control the natural world.

______________________ 2. *True or False?* One of the earliest uses of technology was to harness natural forces.

3. What is *futuring*?

____

____

____

4. Name the five distinct features that futuring emphasizes.

____

____

____

____

______________________ 5. Futuring requires two types of ______—divergent and convergent.

______ 6. Which of the following types of futures is concerned with the types of relationships people want with each other?

A. Biological future
B. Human-psyche future
C. Social future
D. Technological future

______________________ 7. The type of future that is concerned with the types of plant and animal life that will exist is called the ______ future.

8. Identify three widely discussed issues for which technology presents both challenges and promises.

____

____

____

______________________________ 9. Engineers are increasingly pressed to consider ______ use in the technologies they create.

______________________________ 10. The supply of nonreneweable resources is ______—there is a limited quantity of the resource available.

______________________________ 11. *True or False?* It takes time to develop the technology to fully use inexhaustible energy resources.

______________________________ 12. The natural environment has a(n) ______ effect on the safety and health of people.

13. What are the three important forces contributing to the environmental crisis?

____

____

______________________________ 14. The current world ______ is approximately 7.5 billion people, and it is predicted to increase to approximately 9 billion by the year 2044.

15. What is an *environmentalist*?

____

____

____

______________________________ 16. *True or False?* Pollution is most often a product of human activity.

______ 17. Increased UV light and levels of carbon dioxide and other gases causes Earth to retain more heat, which creates a problem called ______.

A. environmentalism
B. the greenhouse effect
C. aquifer
D. pollution

______________________________ 18. An issue directly related to society and ______ is distribution of wealth and industrial power.

______ 19. Which of the following is *not* an action a country should take to overcome a loss of economic power?

A. Change management styles
B. Decrease use of computers
C. Produce world-class products
D. Use flexible, automated manufacturing systems

______________________________ 20. Automated production lines called ______ systems are replacing traditional manufacturing lines.

Name ______________________________ Date __________ Class ________________

CHAPTER 30

Technology and Engineering: A Personal View

Complete the following questions and problems after carefully reading the corresponding textbook chapter.

__________________ 1. *True or False?* A lifestyle is what a person does with business and family life.

2. What were the transportations systems during America's colonial period?

____

____

____

__________________ 3. A(n) ______ is a collection of people who generally live in a community, region, or nation and share common customs and laws.

______ 4. Which of the following is *not* a feature of a central manufacturing operation?
   A. Continuous-manufacturing techniques
   B. Custom parts
   C. Division of labor
   D. Professional management

__________________ 5. The computer allowed the development of ______ manufacturing.

______ 6. Workers in the information age must be willing to ______.
   A. accept job and career changes several times during their work lives
   B. continuously pursue additional education and training
   C. exercise leadership, be self-driven, and accept responsibility for their work
   D. All of the above.

__________________ 7. *True or False?* A technician is a person who processes materials and makes products in manufacturing companies, erects structures, and operates transportation vehicles.

_______________ 8. ______ are people-oriented leaders who also have technical knowledge.

_______________ 9. Whenever possible, a job should match your ______ and abilities.

10. List the six major groups of general job skills.

____

____

____

____

____

_______________ 11. *True or False?* Thinking skills are the ability to use mental processes to address problems and issues.

_______________ 12. Information skills are the abilities to ______, select, and use information.

______ 13. ______ skills are the skills that involve understanding the implications of actions on people, society, and the environment.

A. Thinking
B. Socioethical
C. People
D. Personal

14. What are *ethical skills*?

____

____

____

____

_______________ 15. Teamwork and diversity are examples of ______ skills.

______ 16. Which of the following is *not* a personal skill?

A. Goals
B. Self-learning
C. Self-management
D. Creative thinking

_______________ 17. People must comprehend the political and economic systems directing the development and implementation of ______.

18. List two actions involved in the role of the consumer.

____

____

____

____

_______________________ 19. *True or False?* Activists use public opinion to shape practices and societal values.

______ 20. Which of the following is *not* a major concern involving technology?
A. Energy use
B. Genetic engineering
C. Nuclear power and nuclear-waste disposal
D. Technological employment

_______________________ 21. The rapid evolution of ______ means that many ideas that seem impossible now will be commonplace in the near future.

Name ______________________________ Date __________ Class __________________

CHAPTER 31

Managing and Organizing a Technological Enterprise

Complete the following questions and problems after carefully reading the corresponding textbook chapter.

__________________ 1. People use complex systems to develop and produce ______.

__________________ 2. *True or False?* A chief executive officer (CEO) is an individual who accepts the financial risks of creating a small business and focuses on what customers value in order to develop systems and products to meet desires and expectations.

3. What is *intrapreneurship*?

____

____

____

______ 4. Which of the following is *not* a function of management?
A. Controlling
B. Financing
C. Organizing
D. Planning

__________________ 5. Management involves authority and ______.

6. What is *actuating*?

____

____

____

__________________ 7. *True or False?* A president or chief executive officer is the top manager in a company.

______ 8. Which of the following levels of management is responsible for the entire company's operation?
A. Bottom management
B. Middle management
C. Operating management
D. Top management

______________________ 9. The level of management in a company that is below the president and vice presidents, but above operating management, is called ______ management.

______________________ 10. *True or False?* A vice president directly oversees specific operations in the company and is closest to the people who produce the company's products and services.

11. What is the role of *operating management*?

____

____

____

____

____

____

______________________ 12. Everyone involved with a company is subject to risks and ______.

______ 13. Which of the following is *not* an important feature involved in the formation of a company?
A. Controlling inventory
B. Establishing the enterprise
C. Securing financing
D. Selecting a type of ownership

______________________ 14. *True or False?* A public enterprise is controlled by the government or a special form of corporation; is operated for the general welfare of society; and cannot, or should not, make a profit.

15. What is a *private enterprise*?

____

____

____

____

____

_______ 16. Which of the following is a disadvantage in which the proprietor of a business is responsible for all the debts the business incurs?
A. Debt financing
B. Equity financing
C. Limited liability
D. Unlimited liability

_______ 17. A business in which investors purchase partial ownership in the form of shares of stock is a _______.
A. corporation
B. partnership
C. proprietorship
D. public enterprise

____________________ 18. A periodic payment from a company's profits to investor-owners is called a(n) _______.

____________________ 19. *True or False?* Limited liability is a feature in which an owner's loss is limited to the amount of money he or she has invested.

Match the following terms to the corresponding description or definition.

A. Partnership
B. Proprietorship
C. Corporation

_______ 20. Business in which investors have purchased partial ownership

_______ 21. Business in which a single owner has complete control of a company

_______ 22. Business owned and operated by two or more people

____________________ 23. Once the type of ownership is selected, the company must be _______.

24. What is a *charter*?

____

____

____

25. List four pieces of information, besides the information contained in the charter, that are included in a set of bylaws.

____

____

____

____

Match each term to its description below.

A. Bylaw
B. Charter
C. Board of directors
D. Inside directors
E. Outside directors

_______ 26. People elected to represent the interests of the stockholders

_______ 27. Top managers of a company

_______ 28. A general rule under which a corporation operates

_______ 29. Board of directors members who are not involved in day-to-day operations

_______ 30. "Birth certificate" of a corporation

_______________________ 31. *True or False?* The two basic methods of raising operating funds are equity financing and debt financing.

_______________________ 32. A method for raising money by borrowing it from financial institutions or private investors is called _______ financing.

_______________________ 33. *True or False?* The prime interest rate is a high interest rate.

34. What is a *bond*?

____

____

____

Name ______________________ Date __________ Class ______________

CHAPTER 32

Operating a Technological Enterprise

Complete the following questions and problems after carefully reading the corresponding textbook chapter.

______ 1. Society is made up of major parts called ______.
A. economies
B. enterprises
C. industries
D. institutions

2. List the five basic institutions within society.

____

____

____

______________________ 3. *True or False?* In a free enterprise economic system, the government greatly impacts production and prices.

4. What is *industry*?

____

____

____

____

______________________ 5. Research that seeks knowledge for its own sake is called ______ research.

______________________ 6. *True or False?* Applied research is research that seeks to reach a commercial goal by selecting, applying, and adapting knowledge gathered during basic research.

Match the following areas of industrial activity to the corresponding description.

A. Financial affairs
B. Industrial relations
C. Marketing
D. Production
E. Research and development

_______ 7. Find and create new or improved products and processes

_______ 8. Develop an efficient workforce and maintain positive relations with workers and the public

_______ 9. Develop methods for producing products or services and the desired outputs

_______ 10. Obtain, account for, and pay out funds

_______ 11. Encourage the flow of goods and services from producer to consumer

_______ 12. Which of the following are *not* included in a set of engineering drawings?
A. Architectural drawings
B. Assembly drawings
C. Detail drawings
D. Systems drawings

13. What is an *architectural drawing*?

____

____

____

_______________________ 14. A document that contains the quantities, types, and sizes of materials and hardware needed to build a product or structure is called a(n) ______.

_______________________ 15. *True or False?* Custom manufacturing is a computer-based manufacturing system combining the advantages of intermittent manufacturing with the advantages of continuous manufacturing.

_______________________ 16. The ______ task determines the sequence of operations needed to complete a particular task and the needs for human, machine, and material resources.

_______ 17. In manufacturing, which of the following involves changing the forms of materials to add to the worth of the materials?
A. Engineering
B. Maintaining quality
C. Planning
D. Producing

____________________ 18. Inspection of representative samples is part of a program called ______.

19. List the four important activities involved in marketing.

____

____

____________________ 20. *True or False?* Advertising may be used to present ideas to promote safety or public health.

21. Name three common distribution channels for consumer products.

____

____

22. Name the three main programs of industrial relations activities.

____

____

____

____________________ 23. *True or False?* Labor relations programs are a company's programs that deal with the employees' labor unions.

____________________ 24. Employees who are fully qualified to work are the result of managed employee-______ activities.

______ 25. In ______ training, basic information about the company and its rules and policies are provided to all workers.

A. apprenticeship
B. induction
C. managerial
D. sales

____________________ 26. *True or False?* Simple skills are generally taught through classroom training.

27. What are the two ways in which companies recognize and reward people?

____

____

____

____

____

______ 28. A set rate paid for each hour worked is a ______.

A. benefit
B. commission
C. salary
D. wage

29. What is a *commission*?

____

____

____

______________________ 30. *True or False?* A labor agreement is a contract that establishes pay rates, hours, and working conditions for all employees covered by the contract.

______ 31. Which of the following programs is designed to gain acceptance for company operations and policies?

A. Employee relations
B. Financial relations
C. Labor relations
D. Public relations

______________________ 32. The money a company spends to pay for resources is called ______.

33. What are *retained earnings*?

____

____

____

______________________ 34. The industry-consumer product cycle continues with a constant array of new products being developed and ______ products disappearing.

Name ______________________ Date __________ Class ______________

CHAPTER 33 Understanding and Assessing the Impact of Technology

Complete the following questions and problems after carefully reading the corresponding textbook chapter.

1. List the two major technological actions involved with the use and impacts of technology.

   ____

   ____

_______ 2. Which of the following involves selecting and operating tools, machines, and systems to modify or control the environment?
   A. Assessing technology
   B. Repairing technology
   C. Servicing technology
   D. Using technology

_______ 3. Which of the following is an output that people use, but do not own?
   A. Technological assessment
   B. Technological product
   C. Technological repair
   D. Technological service

______________________ 4. The tension between positive and negative impacts on people or the _______ means that people should use technology wisely.

______________________ 5. *True or False?* When we use technological services, we complete a series of tasks.

6. List the five steps for using technological products.

____

____

____

____

____

______________ 7. Trial and error is an inefficient and expensive approach to ______ an appropriate product.

______________ 8. Once a product has been selected and purchased, the new owner must learn how to ______ the device.

9. List three pieces of information an owner might have to learn to use a new product.

____

____

____

______________ 10. *True or False?* Repairing is doing routine tasks that keep a product operating.

______________ 11. ______ involves replacing broken and worn parts.

______ 12. Options for disposing of products include ______.
A. donating them
B. recycling them
C. sending them to a landfill
D. All of the above.

______________ 13. We can only select, use, and ______ a technological service.

______________ 14. *True or False?* The select-use-evaluate model can be used for any technological service.

______________ 15. An evaluation of the impacts of technology on people, society, and the environment by groups of people is called a ______.

______________ 16. *True or False?* Technology assessment places the task of controlling technology on the government and the courts.

Indicate the order of steps for a technology assessment.

A. First step
B. Second step
C. Third step
D. Fourth step
E. Fifth step

_______ 17. A report is prepared by the assessment group.

_______ 18. Monitor and measure the impact of the technological innovation.

_______ 19. Develop a way to measure the success in meeting the goal.

_______ 20. Review the data and draw conclusions from the measuring and monitoring action.

_______ 21. Identify the underlying goal for the technology.

Name ______________________ Date ____________ Class ______________

ENGINEERING DESIGN CHALLENGE

Temporary Shelter

Read the Engineering Design Challenge *presented at the end of Chapter 4 in the textbook before completing this worksheet.*

1. Identify and define the problem and then write a problem statement (make sure to answer the who, what, where, when, and why).

____

____

____

____

2. Read the Desired Outcomes for this *Design Challenge*. Identify the criteria and constraints which must be taken into consideration when you seek to respond to the problem statement. List them here:

____

____

____

____

____

____

3. Sketch at least two possible solutions to the design problem of a temporary shelter in the following grid. In addition, gather at least two pieces of information (e.g., current solutions, concepts, important details) and record them in the grid.

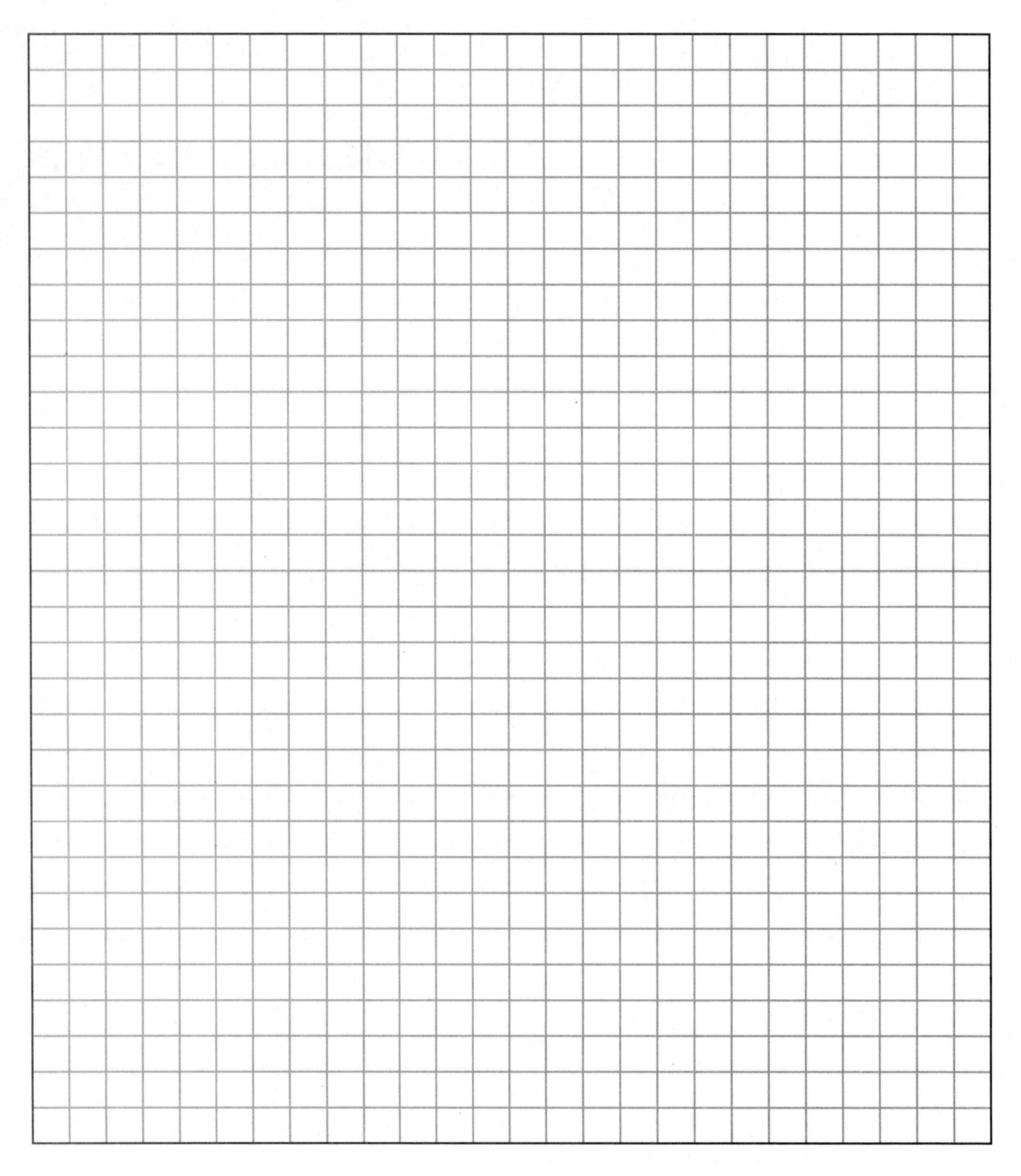

4. Identify one idea for the temporary shelter and further sketch out a proposed solution using this idea in the following grid. Seek feedback, suggestions, and questions from others and record these on the lines provided.

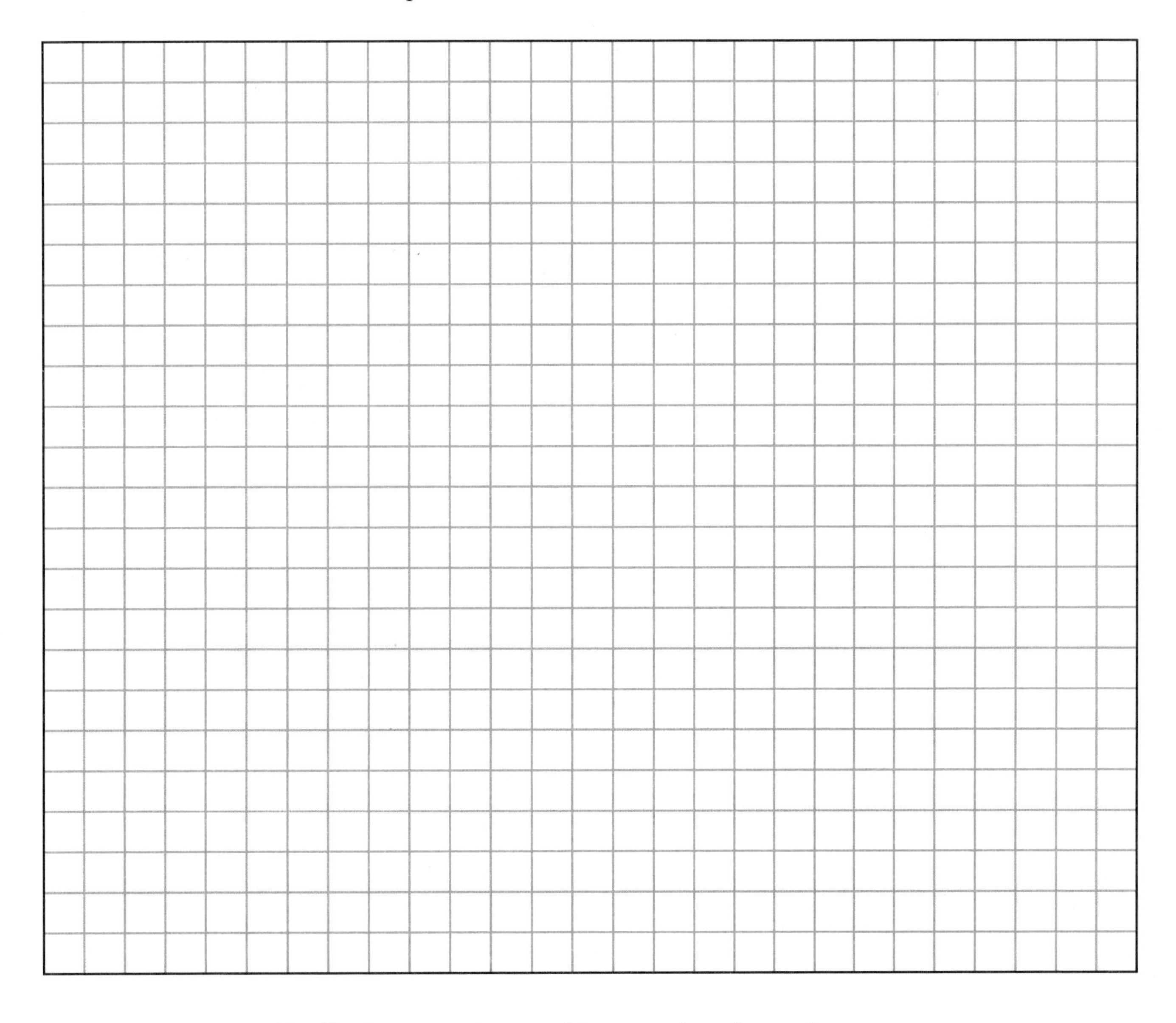

Feedback, suggestions, and key questions from others:

____

____

____

____

____

____

____

5. Using the problem statement, the criteria and constraints, your ideas, and the feedback from others, develop your final solution for a temporary shelter. Sketch your final solution in the grid that follows. If you have access to design software, develop your design idea using the software.

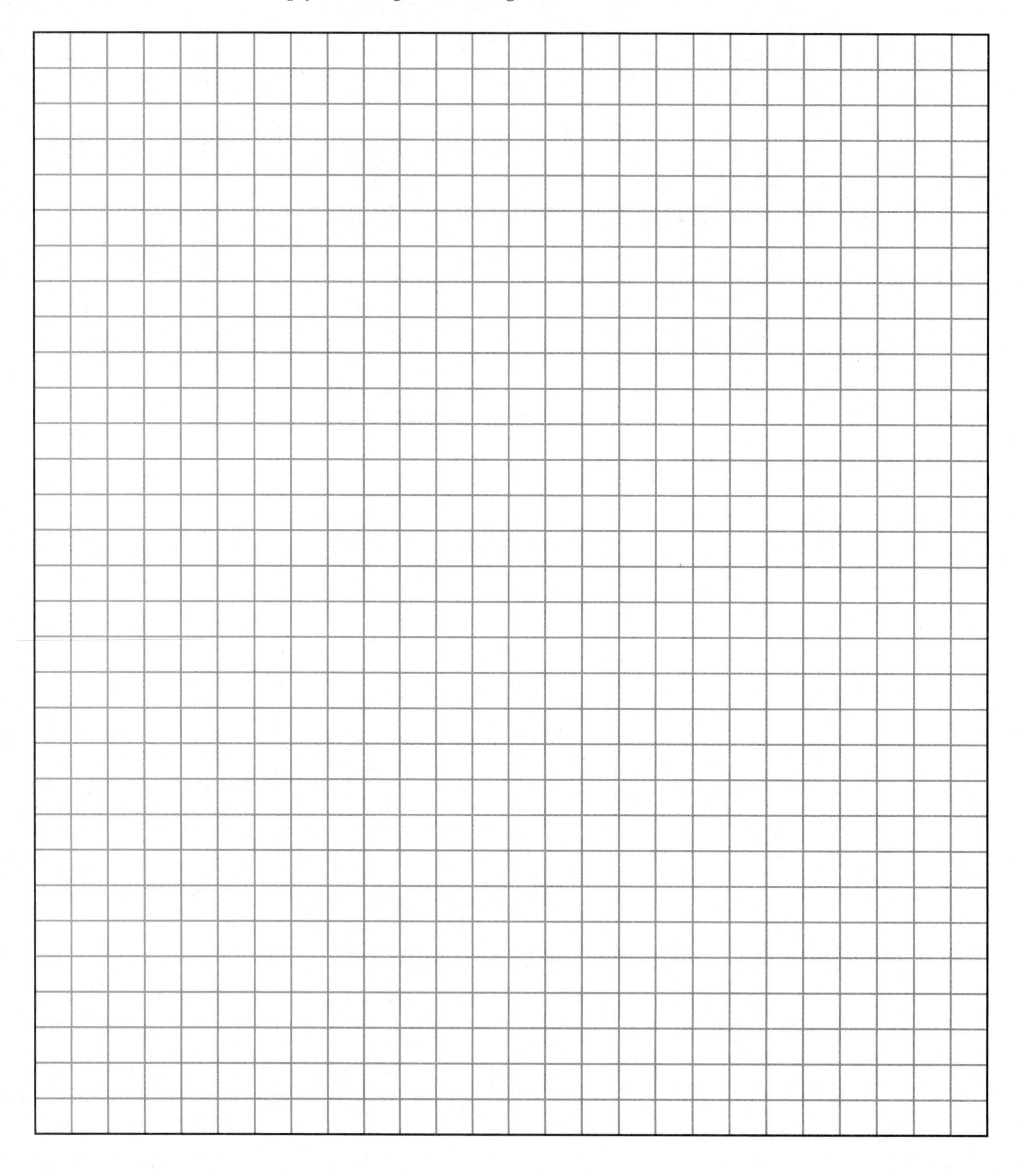

Name ______________________ Date __________ Class ______________

ENGINEERING DESIGN CHALLENGE

Design Problem

Read the Engineering Design Challenge *presented at the end of Chapter 5 in the textbook before completing this worksheet.*

1. Write a definition of the design problem before your group.

____

____

____

____

2. List points other members of your group presented that are not stated in your definition.

____

____

____

____

____

____

3. Write the definition your group has developed using the several individual definitions the group's members presented.

____

____

____

____

4. List at least three criteria your product must meet for each category on this sheet.

Engineering criteria

____

____

____

____

Production criteria

____

____

____

____

Marketing criteria

____

____

____

____

Human criteria

____

____

____

____

Financial criteria

____

____

____

____

Environmental criteria

____

____

____

____

Name ____________________________________

5. Develop at least three rough sketches of a product fitting your company's product profile. Each sketch should be a complete solution of the product need.

6. Develop a refined sketch with dimension of the best idea you have for a product fitting your company's product file.

Name ___________________________

7. Make a detailed drawing of one part of the product your group has chosen. Make sure the design meets the product definition and criteria.

Product name: ______________________	Drafter: ______________ Class: ______
Part name: ______________________	Scale: ______________ Date: ______

8. Produce a bill of materials for the product your group has developed.

Bill of Materials

Product: ______________ Product Development Group Number: ______________

Part Number	Quantity Needed	Part Name	Size			Material
			Thickness	Width	Length	

Name ______________________ Date __________ Class ______________

ENGINEERING DESIGN CHALLENGE

Making a Board Game

Read the Engineering Design Challenge *presented at the end of Chapter 6 in the textbook before completing this worksheet.*

1. Watch your teacher demonstrate how to dimension a sketch. On the following grid, develop a dimensioned sketch of the product you have selected to build.

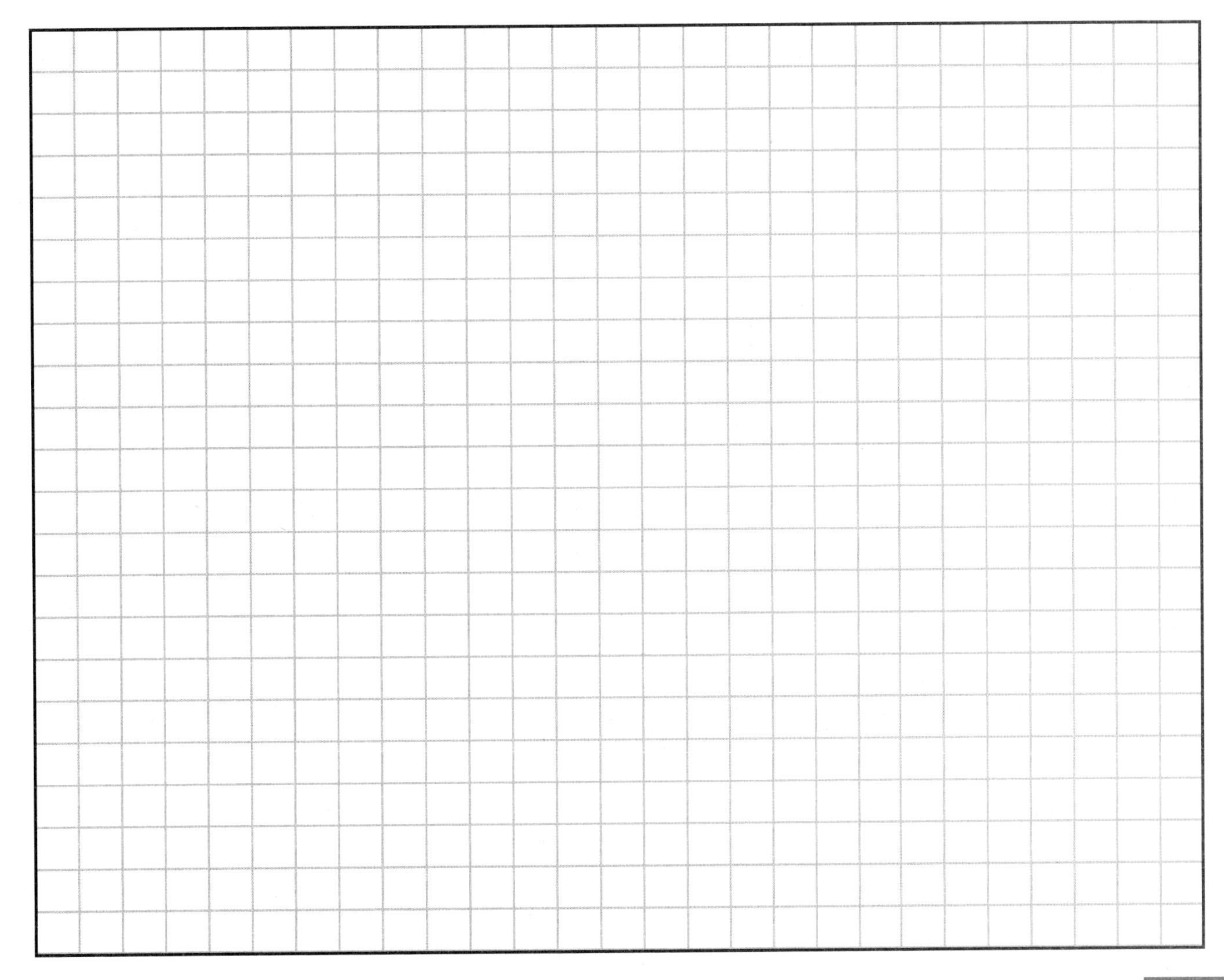

2. As you watch the demonstration of the procedure for making the product, complete the following chart.

Step #	Task	Machine or Tool	Safety Precautions and Considerations

(Chart Continued)

Name ________________________________

Step #	Task	Machine or Tool	Safety Precautions and Considerations

Name ______________________ Date __________ Class ______________

ENGINEERING DESIGN CHALLENGE

Bookend Design

Read the Engineering Design Challenge *presented at the end of Chapter 8 in the textbook before completing this worksheet.*

1. Write a definition of the design problem before your group.

____

____

____

____

2. List points other members of your group presented that are not stated in your definition.

____

____

____

____

____

____

3. Write the definition your group has developed using the several individual definitions the group's members presented.

____

____

____

____

4. List at least three criteria your product must meet for each category on this sheet.

Engineering criteria

Production criteria

Marketing criteria

Human criteria

Financial criteria

Environmental criteria

Name ________________________________

5. Develop at least three rough sketches of a product fitting your company's product profile. Each sketch should be a complete solution of the product need.

6. Develop a refined sketch with dimension of the best idea you have for a product fitting your company's product file.

Name ___________________________________

7. Make a detailed drawing of one part of the product your group has chosen. Make sure the design meets the product definition and criteria.

Product name: ____________________	Drafter: ____________________ Class: ______
Part name: ____________________	Scale: ____________________ Date: ________

8. Produce a bill of materials for the product your group has developed.

Bill of Materials

Product: ____________________ Product Development Group Number: ____________________

Part Number	Quantity Needed	Part Name	Size			Material
			Thickness	Width	Length	

Name ______________________ Date __________ Class ______________

ENGINEERING DESIGN CHALLENGE

Newtonian System Design

Read the Engineering Design Challenge *presented at the end of Chapter 9 in the textbook before completing this worksheet.*

1. Identify and define the problem and then write a problem statement (make sure to answer the who, what, where, when, and why).

   ____

   ____

   ____

   ____

2. Your design challenge requires you to design a system with at least two steps for each of Newton's laws. The overall objective of the system is to raise a flag and teach others about Newton's laws. Use the following chart to brainstorm and sketch possible ways of representing and teaching about each of Newton's Three Laws in your system.

Newton's Law	Area for brainstorming and sketching possibilities
1. An object at rest will remain at rest and an object in motion will remain in motion at a constant velocity, unless acted on by an external force.	

Continued on next page

Newton's Law	Area for brainstorming and sketching possibilities
2. The force of an object is equal to its mass multiplied by its acceleration.	
3. For every action by a force, there is an equal and opposite reaction by another force.	

3. Work with your classmates to identify which ideas will be used in your system. Write a description for each of the actions in your system and how you will teach others about the Newtonian law for each action. For each action, identify which of Newton's laws this step demonstrates and the supplies necessary to build the system.

Action (e.g., a ball rolls down an inclined plane to knock over a domino)	Newton's Law (1, 2, or 3)
A.	
B.	
C.	
D.	
E.	
F.	

4. Sketch your final system design in the grid below. Label each step in the sequence from the first to the last.

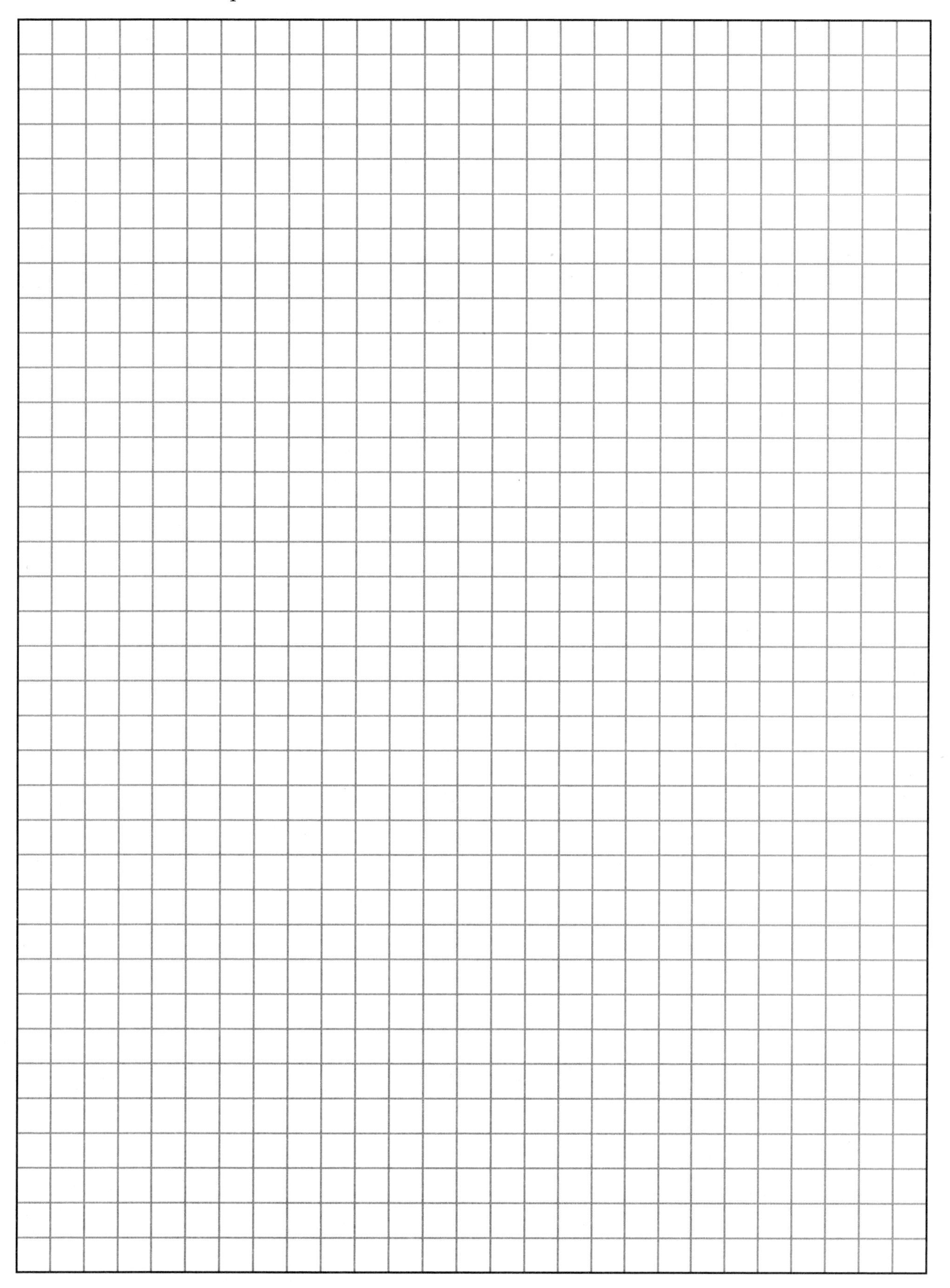

Name ______________________________ Date __________ Class ____________________

ENGINEERING DESIGN CHALLENGE

Simple Machine Science Kit

Read the Engineering Design Challenge presented at the end of Chapter 10 in the textbook before completing this worksheet.

1. List the materials you will tentatively include in your physics kit.

- ______________________________
- ______________________________
- ______________________________
- ______________________________
- ______________________________
- ______________________________
- ______________________________
- ______________________________
- ______________________________
- ______________________________
- ______________________________
- ______________________________

2. On the following grid, sketch two devices that can be built from the list of materials.

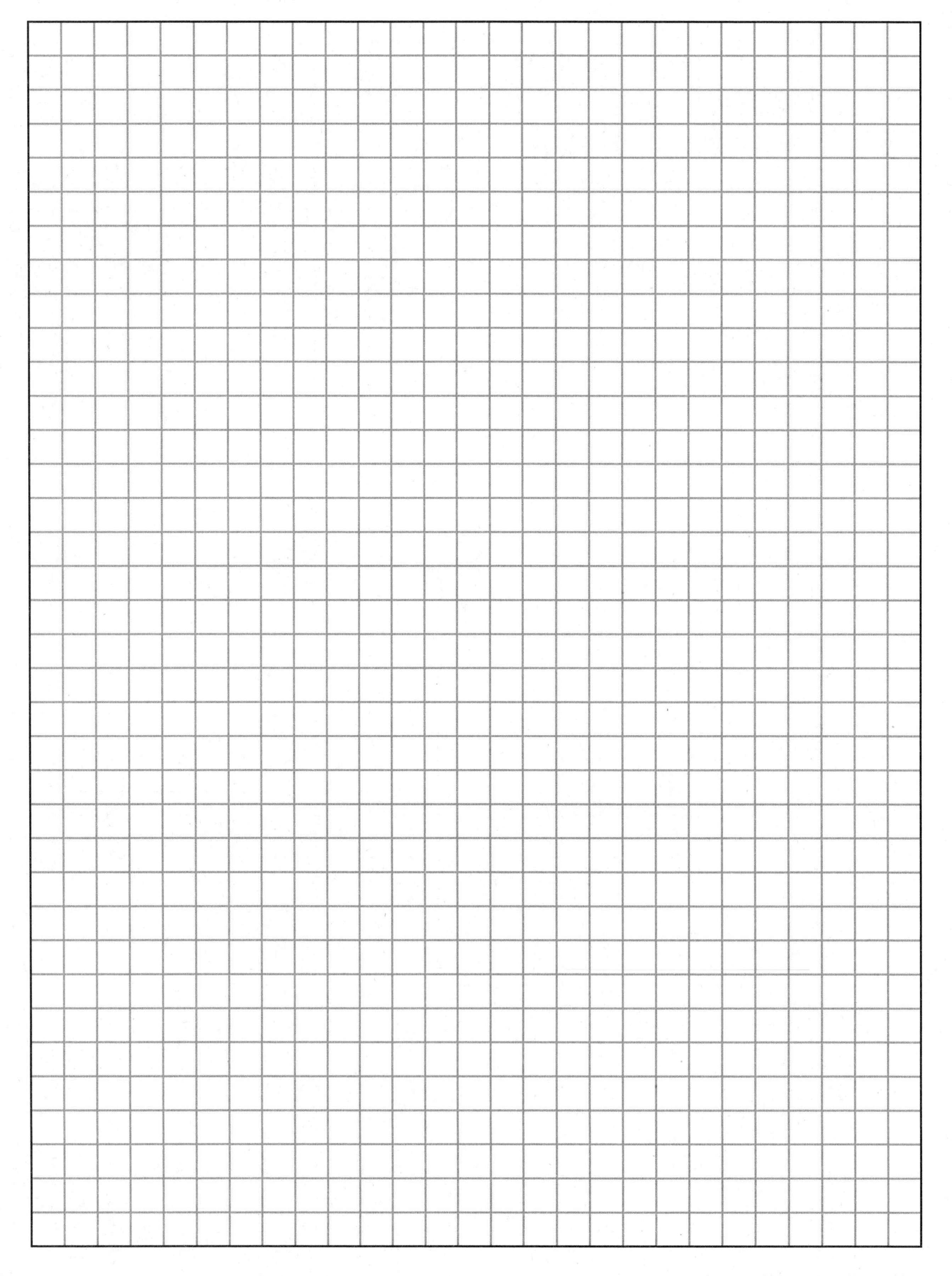

3. Ask members of your group for feedback about your kit and the designs. List their comments, suggestions for improvement, and any questions they asked.

____

____

____

____

____

____

4. List the materials needed for your improved kit.

- ____
- ____
- ____
- ____
- ____
- ____
- ____
- ____
- ____
- ____
- ____
- ____

5. On the following grid, sketch a device that can be built from the improved kit.

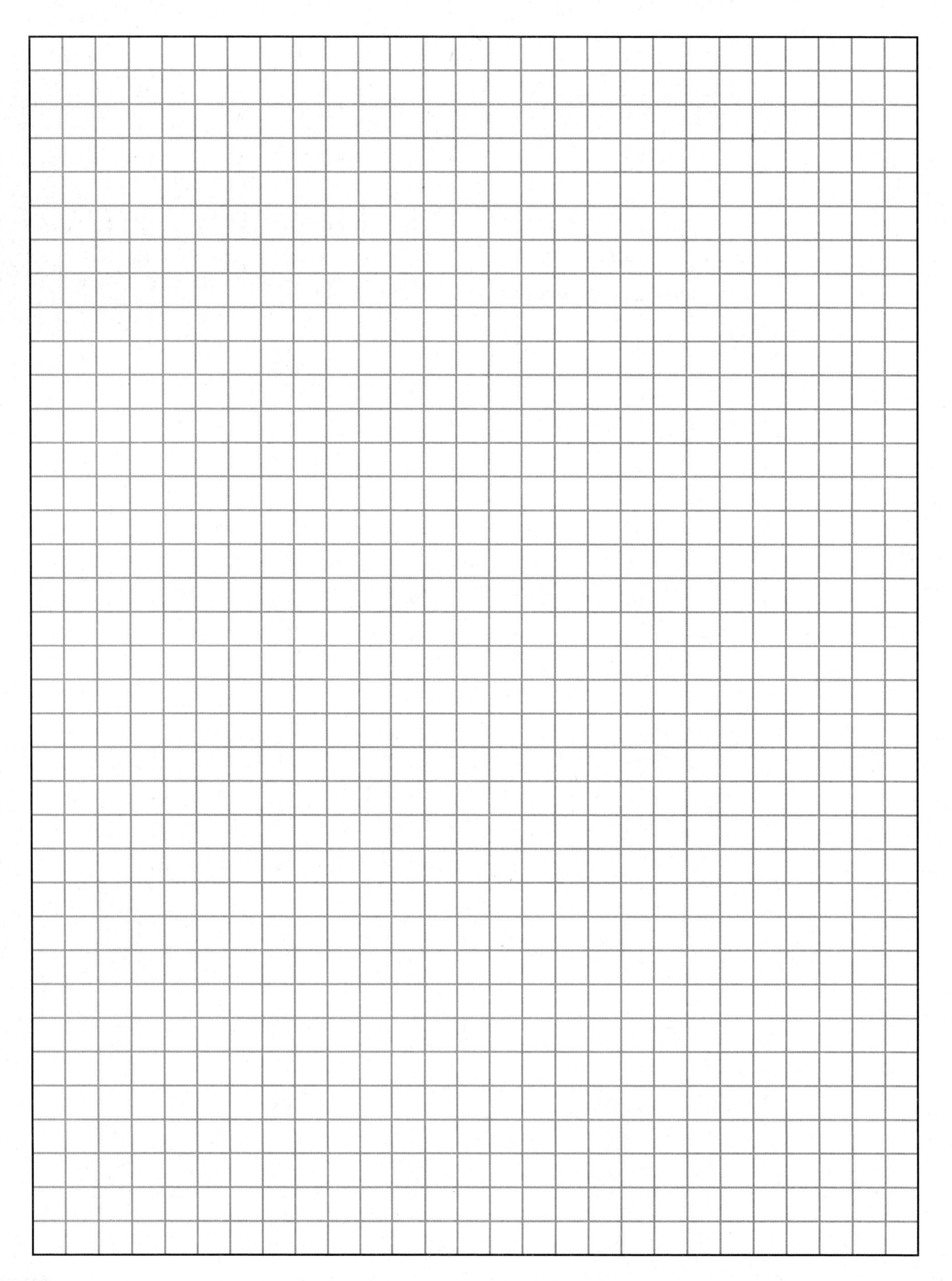

Name ____________________ Date __________ Class ____________

ENGINEERING DESIGN CHALLENGE

Automated Control System Challenge

Read the Engineering Design Challenge *presented at the end of Chapter 12 in the textbook before completing this worksheet.*

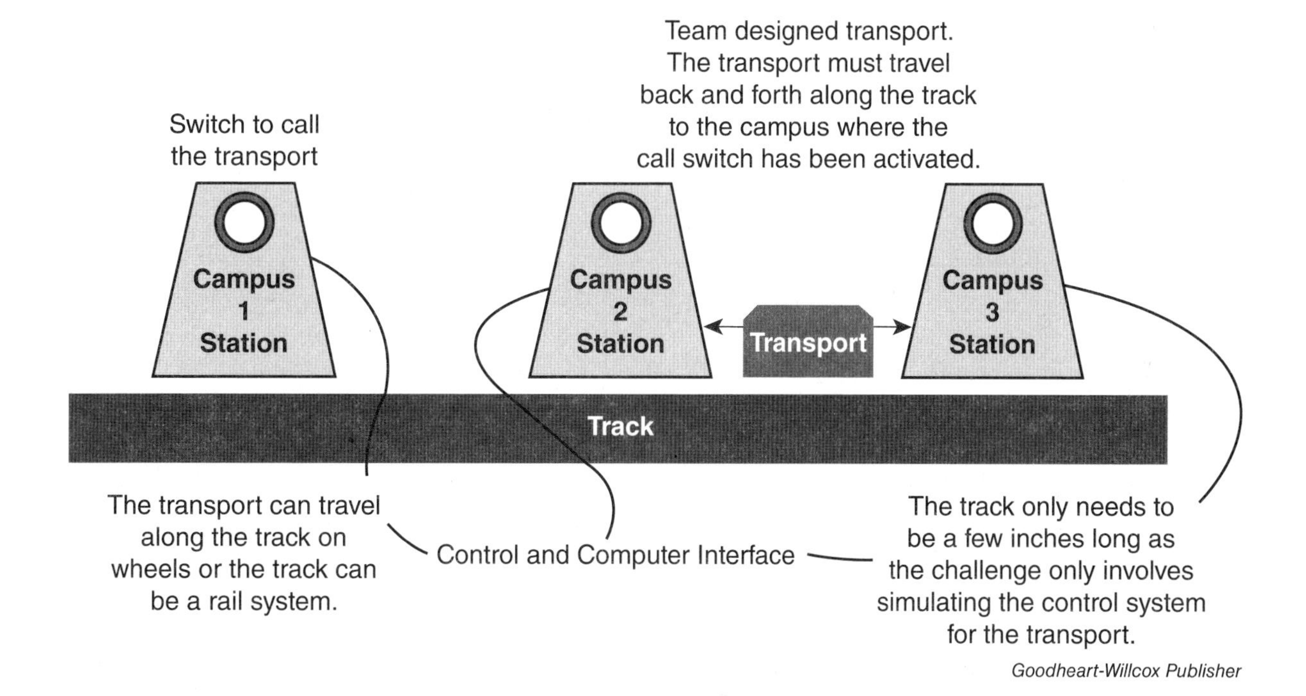

Goodheart-Willcox Publisher

1. Identify the criteria, constraints, and desired outcomes for your solution. Use these to define and construct a problem statement (make sure to answer the who, what, where, when, and why).

____

____

____

____

2. The design challenge specifies that each station will have three switches. Use the table that follows to identify the output associated with each switch.

Station	Switch	Output (action)
Campus 1	1	
	2	
	3	
Campus 2	1	
	2	
	3	
Campus 3	1	
	2	
	3	

3. Your instructor will provide you with robotics equipment and software (e.g., Lego Mindstorms, VEX Robotics, etc.). Work with your classmates to brainstorm, design, and sketch a possible solution of the transportation challenge (remember to include any pertinent wiring, sensors, motors, or other equipment).

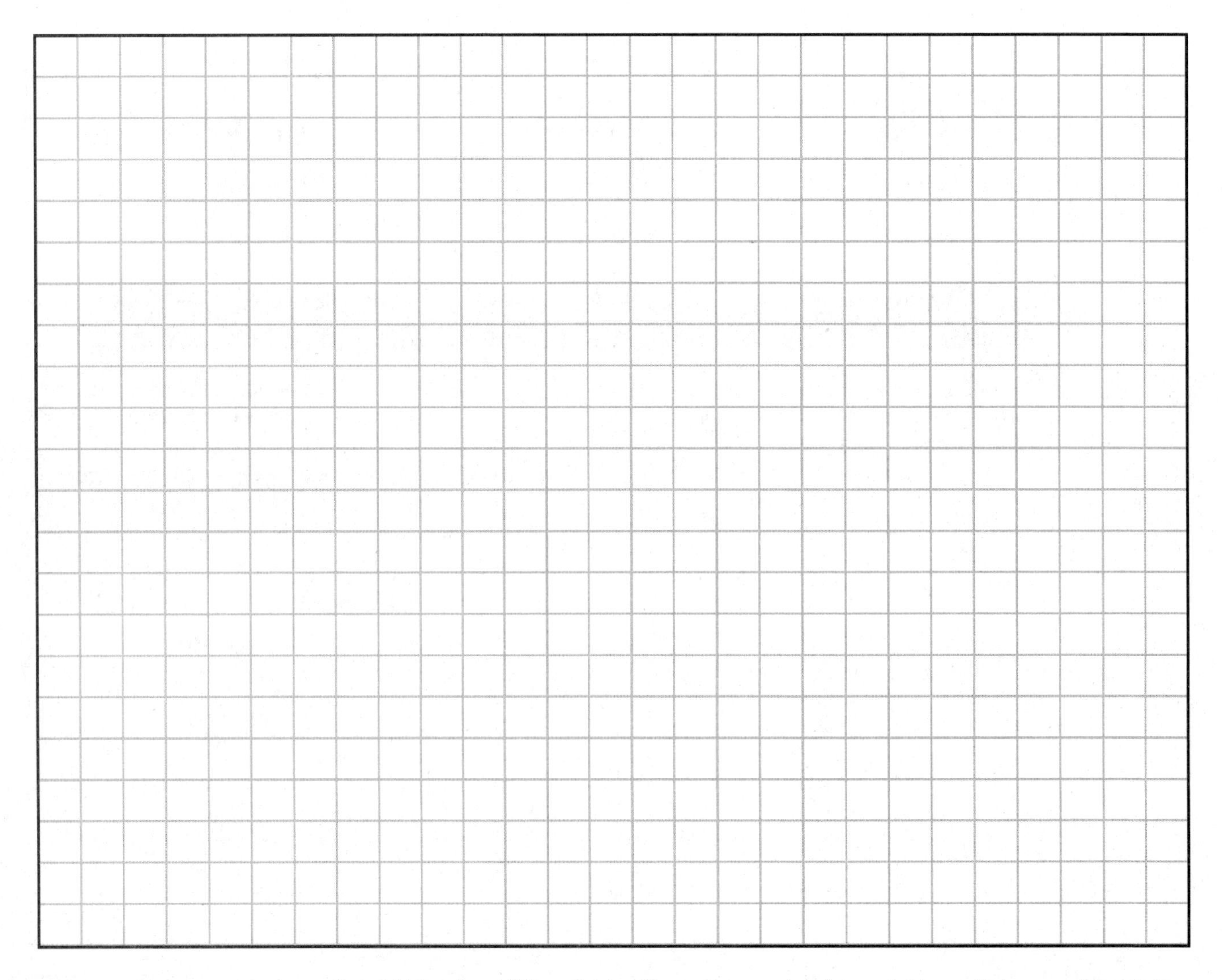

4. Present your solution (in the previous sketch) to your instructor and receive permission to build or program a prototype of your proposed solution. Record the modifications, setbacks, or improvements made to your plans as you build and test your solution.

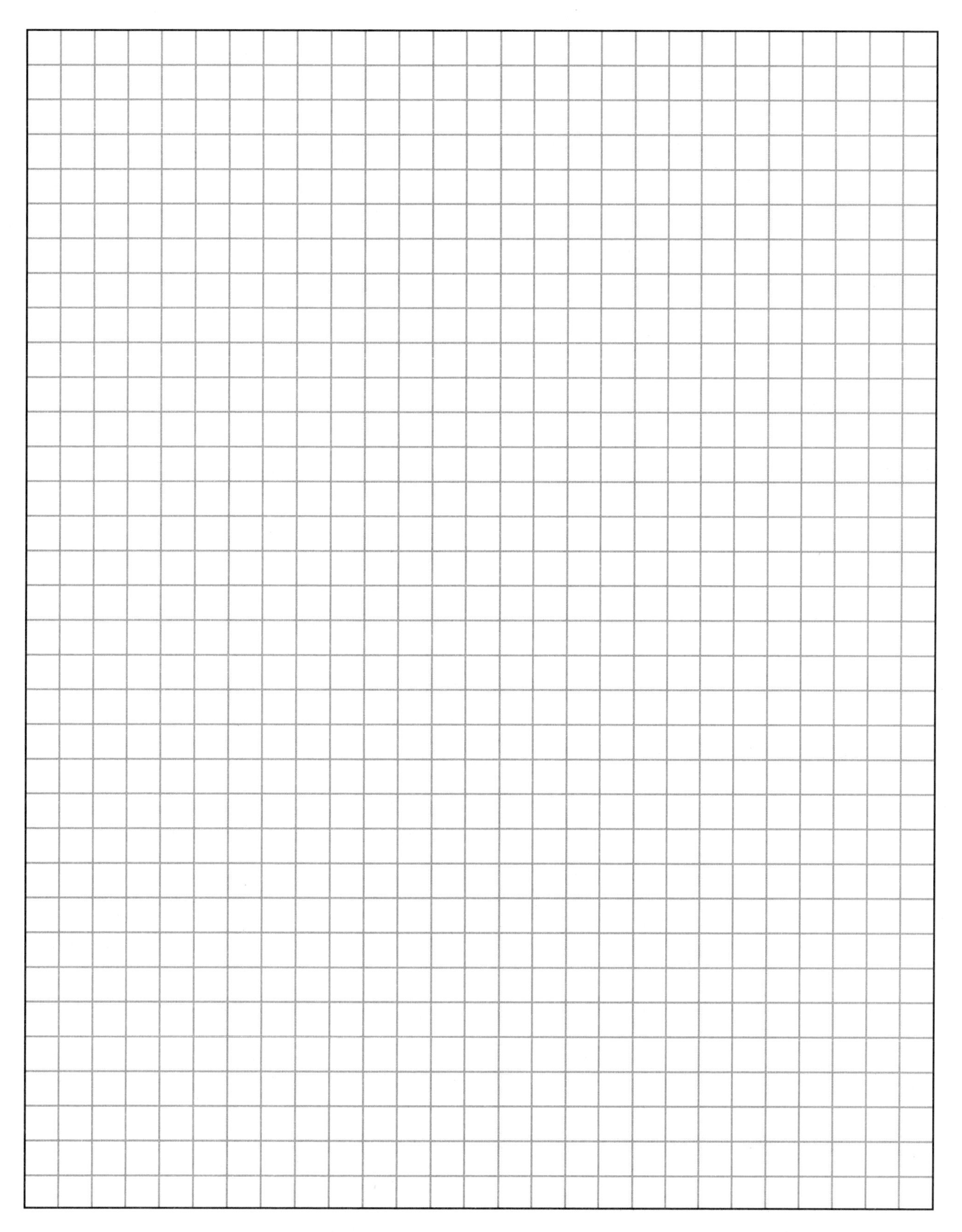

Name ________________ Date ________ Class ____________

ENGINEERING DESIGN CHALLENGE

Lumber Processing System

Read the Engineering Design Challenge *presented at the end of Chapter 14 in the textbook before completing this worksheet.*

1. Identify each of the criteria and constraints outlined in the challenge. List them in the chart that follows. Use the criteria and constraints to form a problem statement. Record this statement in your engineering notebook.

Criteria	Constraints

2. List the necessary actions, equipment, and processes to accomplish each of the required steps (example in first row of table).

Action	Equipment	Process
Cutting logs to length	*Servo motors (2)*	*Motor A will move the log into the cutting position by running for 10 seconds. Motor B will run for 3 seconds and lower the blade into position to cut the log.*

Name ____________________________________

3. Sketch at least two possible solutions to the challenge in the following sketch grid. Present your ideas to your classmates. Ask them for comments and suggestions on areas for possible improvement and refinement. Include their feedback in your drawings.

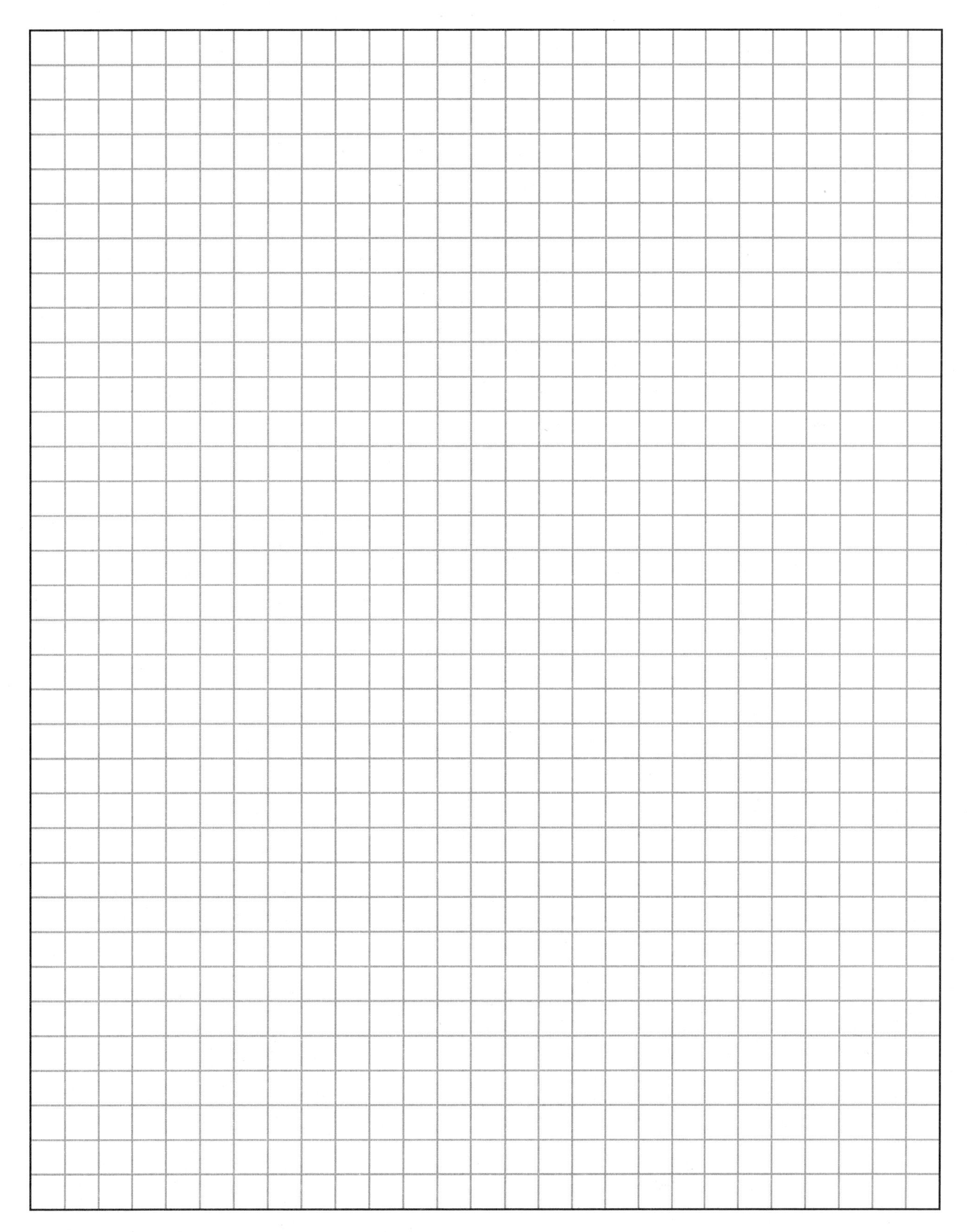

4. Using your sketches, the feedback from your classmates, and your other resources, sketch the solution you will prototype below (be sure to include the necessary motors, sensors, and switches).

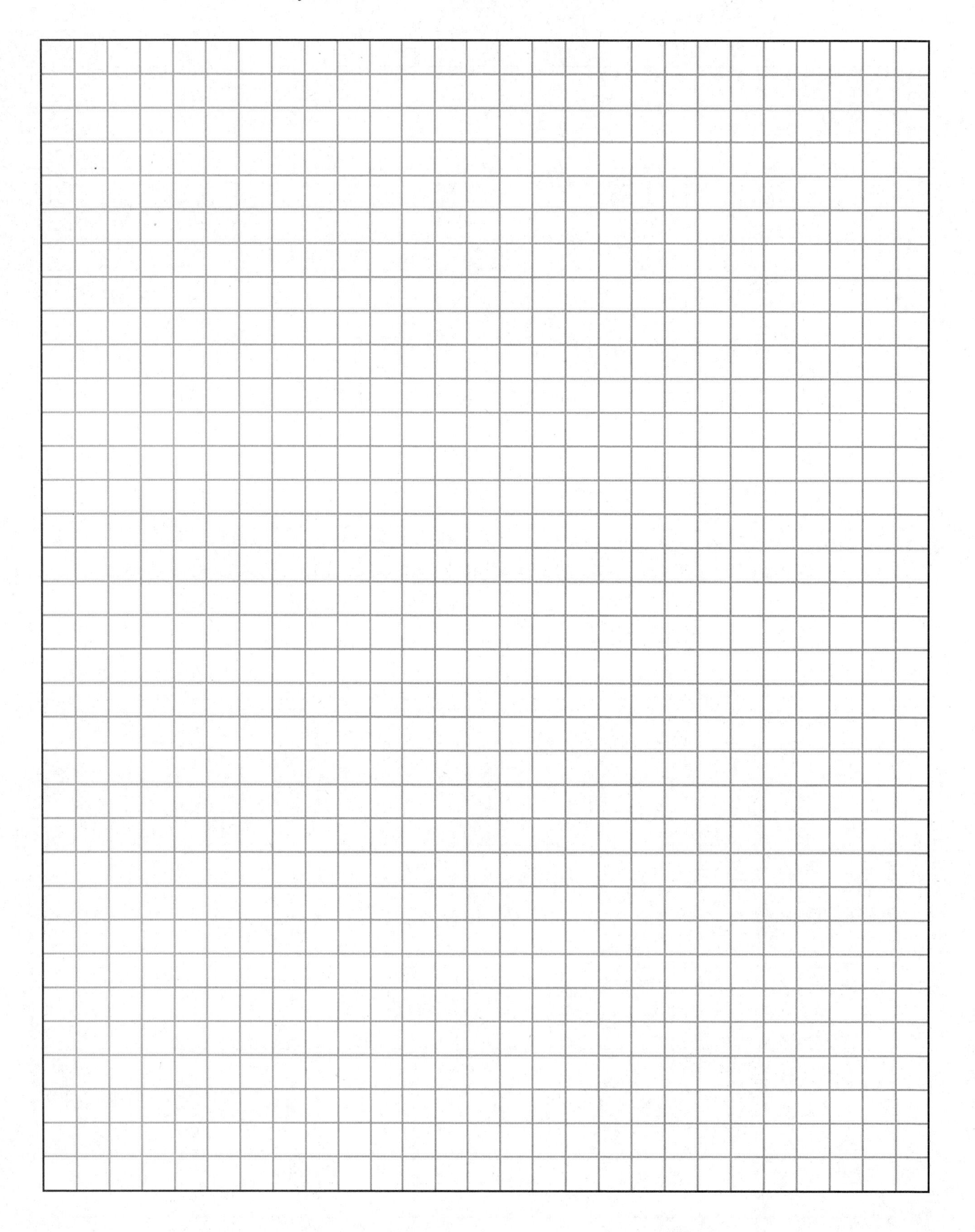

Name ______________________________ Date ___________ Class ___________________

ENGINEERING DESIGN CHALLENGE

Manufacturing Enterprise

Read the Engineering Design Challenge *presented at the end of Chapter 15 in the textbook before completing this worksheet.*

1. Write a complete problem statement for the design challenge.

____

____

____

2. Brainstorm and research several current and past fashion trends that could influence your thinking for this design challenge. Use the prompts to help you in your brainstorming.

Prompt	Your Response
What are some current fashion trends?	
What trends relate to shoes, shirts, pants, hats, backpacks, toys, and accessories?	

Continued on next page

Prompt	Your Response
Ask your parents/guardians to describe fashion trends they can think of.	
Name one thing that all your friends (and you) have had/worn in common.	
What is one thing that was popular when you were younger?	
Ask an older neighbor, sibling, or family member to identify a few things that were popular when they were in school.	

3. Choose one idea to focus on and explain how it will be customizable.

____

____

____

____

____

4. Identify the manufacturing processes that will go into this product:

____

____

____

____

____

5. Use the following space to sketch out the production process for your product. Include the entire process, from original idea to finished merchandise. Refer to the textbook for assistance.

Name ______________________ Date __________ Class ______________

ENGINEERING DESIGN CHALLENGE

Advanced Material Creation

Read the Engineering Design Challenge *presented at the end of Chapter 16 in the textbook before completing this worksheet.*

1. Write a complete problem statement for the design challenge.

____

____

____

____

2. Several criteria for the new material were provided in the design challenge. Using these criteria and the information from the chapter, identify potential materials that could be used to satisfy each criteria (example listed in first column).

Property Criteria	**Materials**						
	Rubber						
Mechanical: Hardness, corrosion resistance, and impact strength	*Low hardness, high anti-corrosion, high impact strength*						
Thermal: Thermal conductivity; thermal fatigue	*Low thermal conductivity, low thermal fatigue*						

Continued on next page

Property Criteria	**Materials**						
Electrical: Low conductivity; high conductivity	*Low conductivity*						
Raw material cost	*Low*						
Processing costs	*Low*						
Machinability	*Low*						
Casting properties	*N/A*						
Tooling required	*Medium*						

3. Identify which materials may be useful in your final design. Circle these materials in the previous table. Sketch a possible mock-up of your design in the sketch grid.

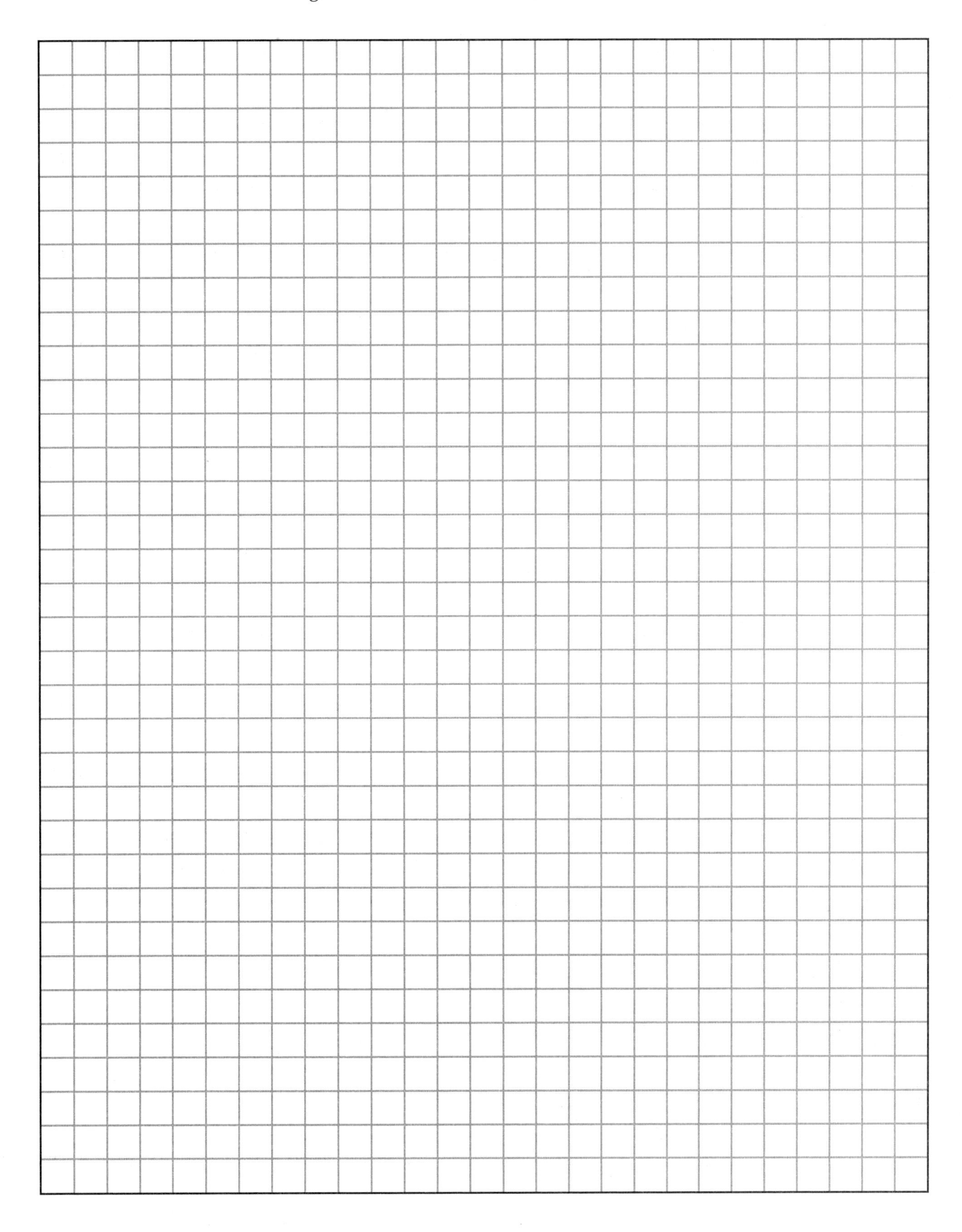

4. Identify the materials of each component of your design (refer to your previous drawing). List the materials in the table that follows. Justify your choice/use of each material.

Material	Why did you choose to use this material?

Name ______________________ Date __________ Class ______________

ENGINEERING DESIGN CHALLENGE

Load Bearing Heavy Engineering Structure

Read the Engineering Design Challenge *presented at the end of Chapter 17 in the textbook before completing this worksheet.*

1. Your challenge is to build a bridge using only limited supplies. The supplies are listed in the table that follows. Identify the strengths and weaknesses of each item.

Item	Strengths	Weaknesses
1/8″ × 1/8″ Balsa wood strip		
Wood glue		
2-lb Test fishing line		

2. Sketch at least three possible bridge designs. Identify the substructure and the deck on each bridge. Present your ideas to classmates and collect their feedback and suggestions for improvement.

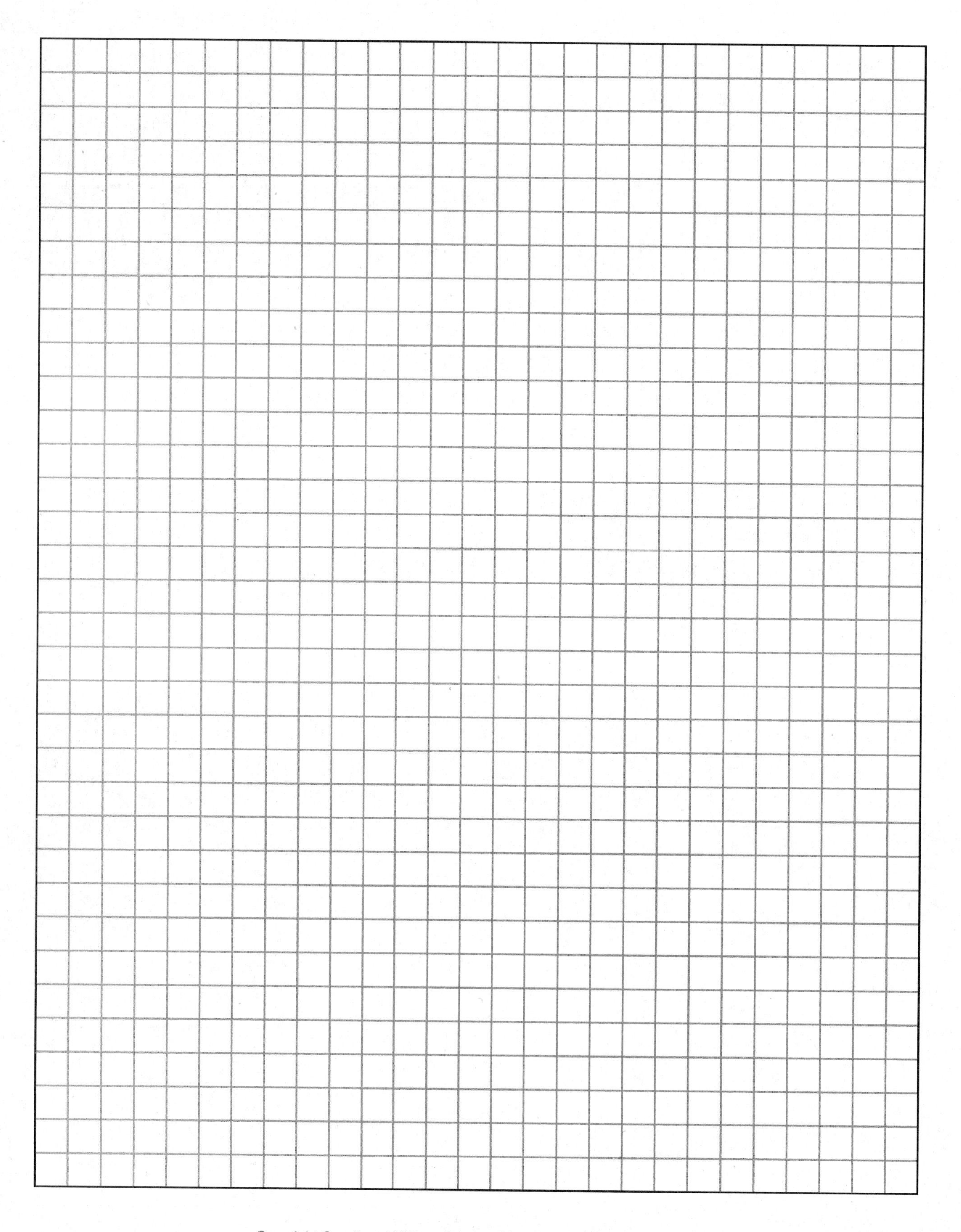

Name ________________________________

3. Sketch your final bridge design. Calculate and record the efficiency of your bridge.

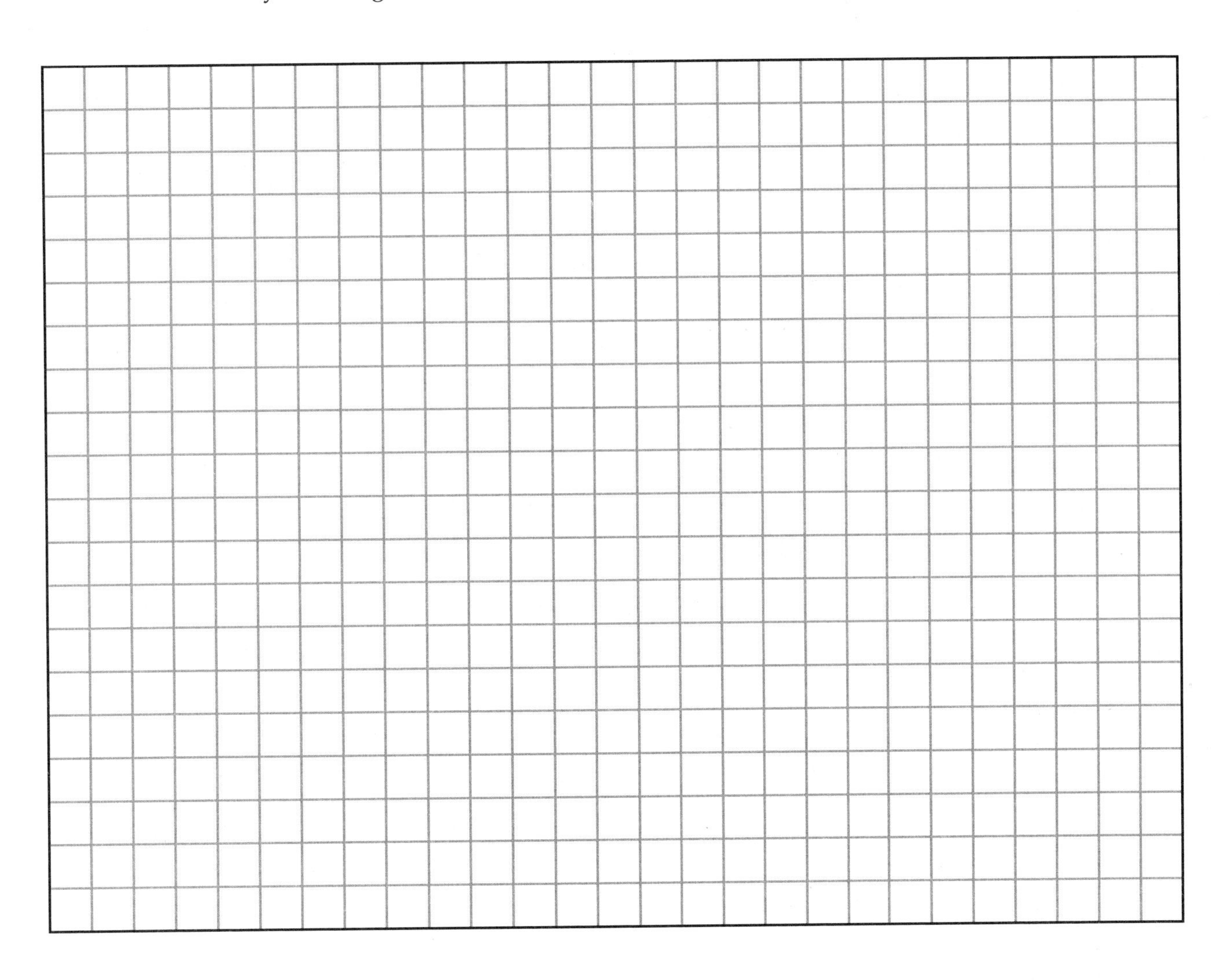

Bridge weight: ______________ Total weight held: ______________

Bridge efficiency: ______________ Bridge efficiency = $\frac{\text{maximum load (g)}}{\text{mass of structure (g)}}$

4. Answer the questions that follow; then use your answers to develop a technical report that describes how your bridge was designed using engineering principles.

   A. How did you decide what type of bridge to use? What are the strengths and weaknesses of the design you chose?

   ____

   ____

   ____

   ____

   ____

B. What materials did you use in the construction of the bridge? What are the strengths/weaknesses of each material? How did you use the strengths to your advantage and overcome the weaknesses of each material?

C. What changes would you make next time based on your experience?

Name ______________________ Date __________ Class ______________

ENGINEERING DESIGN CHALLENGE

Vertical Structure Challenge

Read the Engineering Design Challenge *presented at the end of Chapter 18 in the textbook before completing this worksheet.*

1. Your challenge is to build a tower using only limited supplies. The supplies are listed in the table that follows. Identify the strengths and weaknesses of each item. Pay special attention to the requirements of the earthquake test and the weight the tower must hold.

Item	Strengths	Weaknesses
Copy paper		
3/4″ Cellophane tape		

2. Sketch two possible tower designs. Present your ideas to classmates and collect their feedback and suggestions for improvement. Pick one of the solutions and construct a tower for testing.

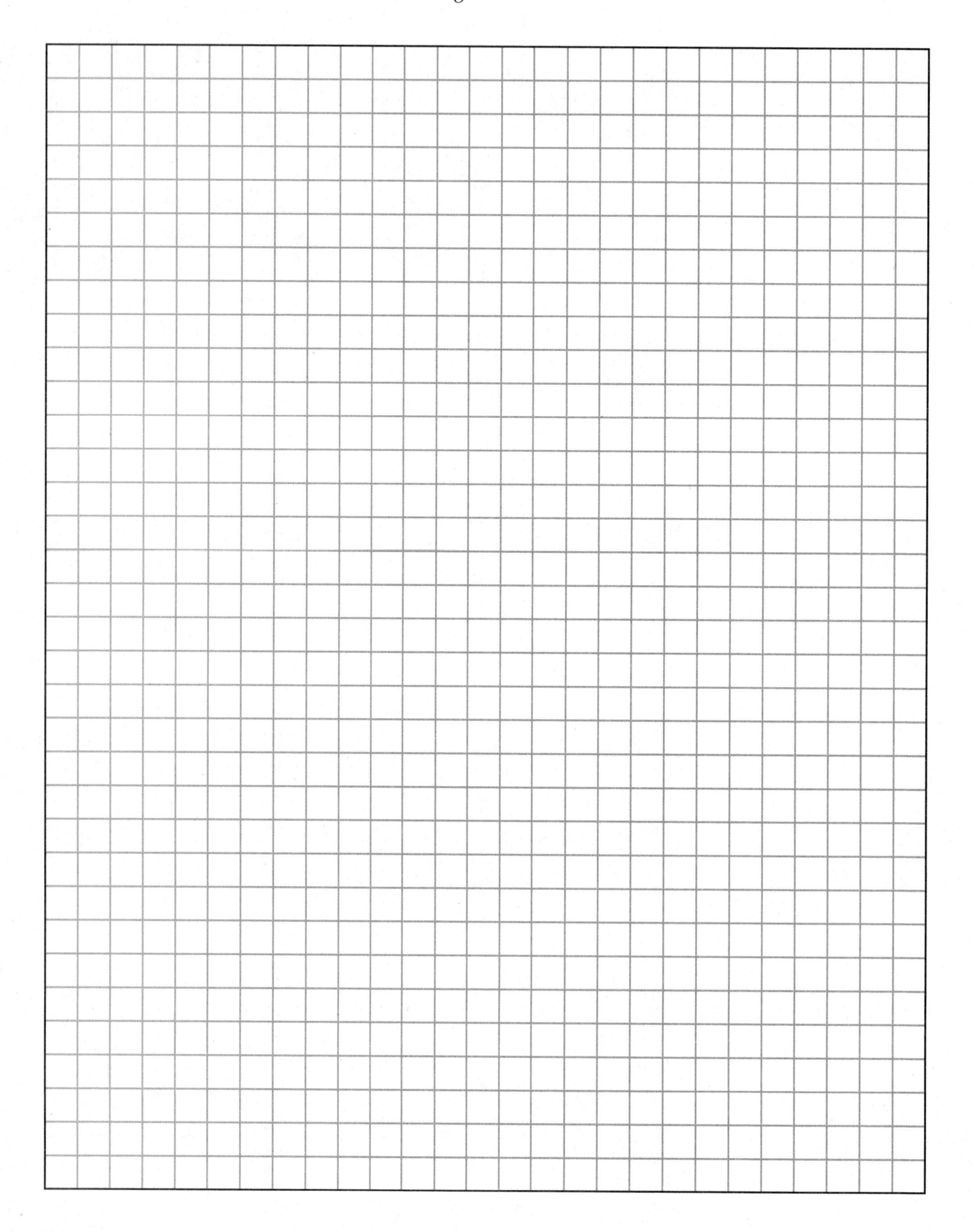

Name ______________________________

3. Test your initial design on the earthquake-testing platform and record the outcome. What happened? Was your design successful? Why or why not? What can you do to improve your tower's design?

____

____

____

____

____

____

____

____

4. Sketch your final tower design. Use the space indicated to record and calculate the efficiency of your tower.

Tower mass: ______________ Total weight held: ______________

Tower height: ______________ Tower efficiency: ______________

5. Answer the questions that follow. Use your answers to develop a technical report that describes how your tower was designed using engineering principles.

   A. How did you decide what type of tower to build? What are the strengths and weaknesses of the design you chose?

   ____

   ____

   ____

   ____

   ____

   ____

   B. How did you use the materials provided in the construction of the tower? What are the strengths/weaknesses of each material? How did you use the strengths to your advantage and overcome the weaknesses of each material?

   ____

   ____

   ____

   ____

   ____

   ____

   C. What changes would you make next time based on your experience?

   ____

   ____

   ____

   ____

   ____

   ____

Name ______________________________ Date __________ Class __________________

ENGINEERING DESIGN CHALLENGE

Wind-Powered Electricity Generator

Read the Engineering Design Challenge *presented at the end of Chapter 20 in the textbook before completing this worksheet.*

1. On the following grid, draw a dimensioned sketch of the energy converter you are going to build.

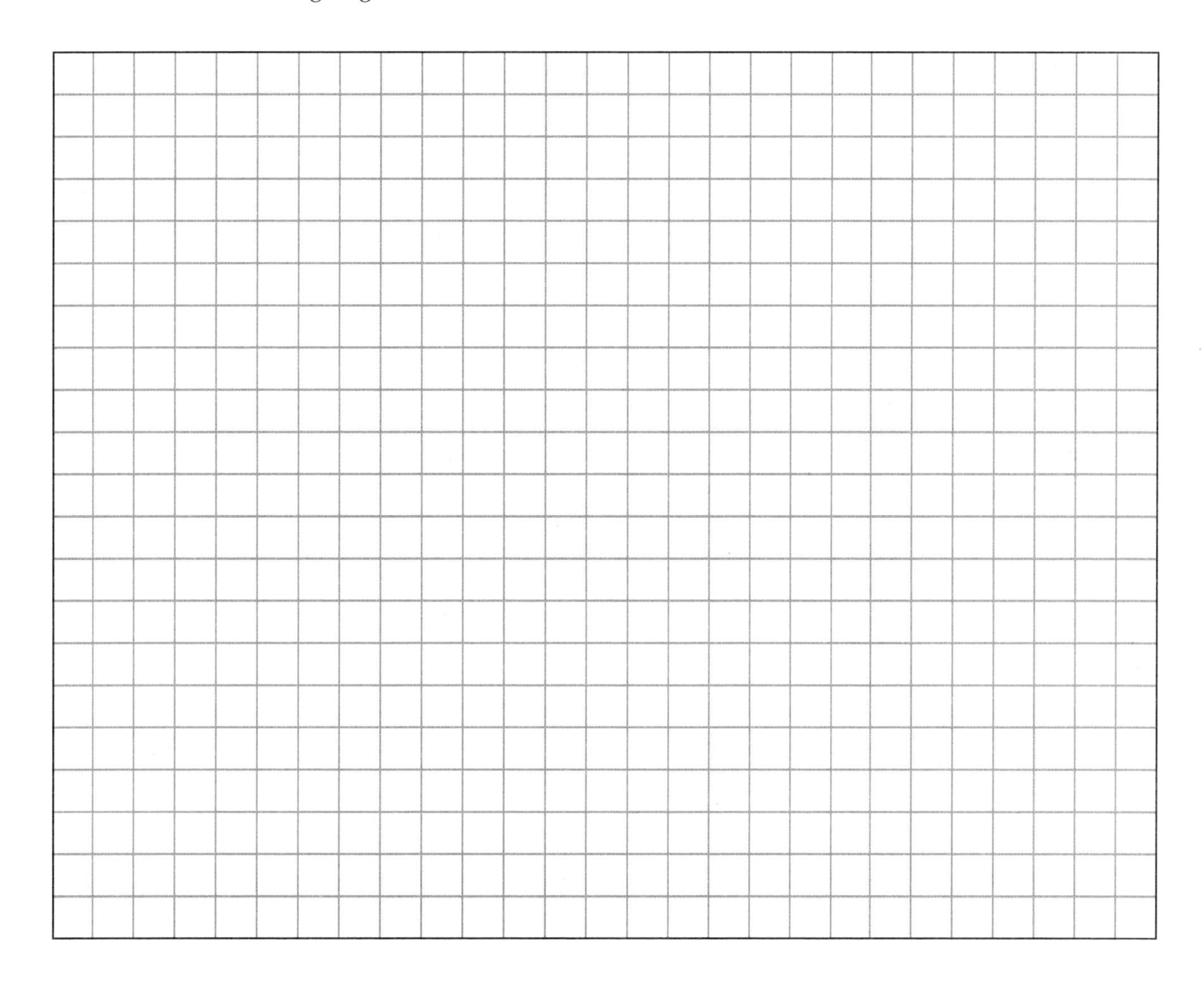

2. Develop a list of materials needed to make your model energy converter.

Quantity	Part Name	Size	Material

3. List the steps you will use to make the parts for your energy converter.

Step with Description	Machine	Safety Considerations

Name ____________________ Date __________ Class ____________

ENGINEERING DESIGN CHALLENGE

Public Service Announcement for Engineering

Read the Engineering Design Challenge *presented at the end of Chapter 21 in the textbook before completing this worksheet.*

Advertising theme: ____________________

Suggested slogans: ____________________

____

____

____

____

____

____

____

____

____

____

____

Rough Sketch #1

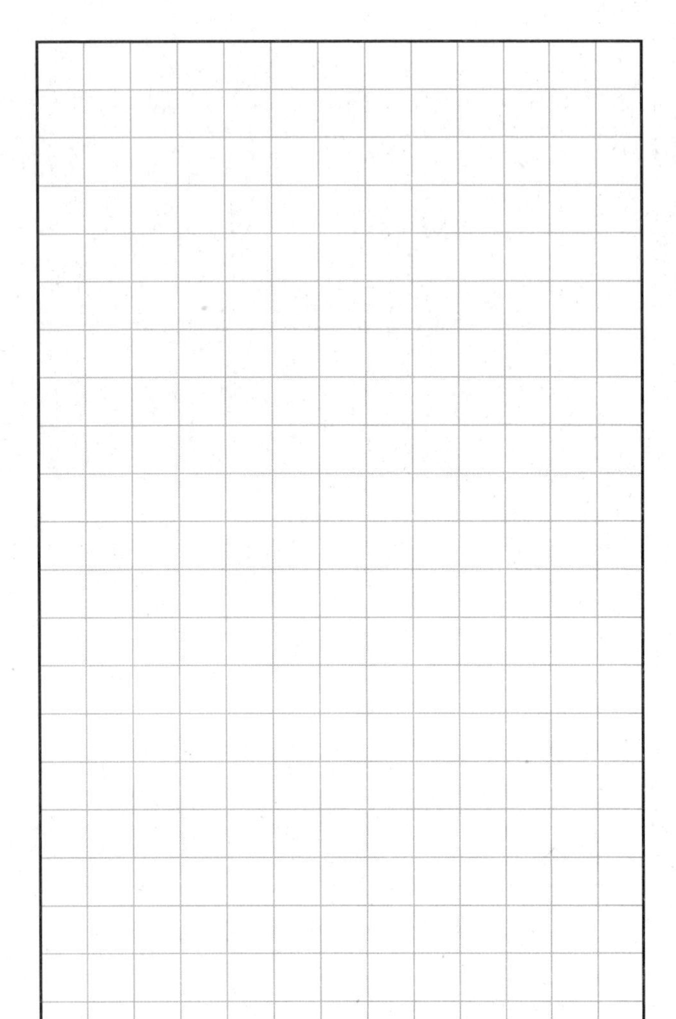

Rough Sketch #2

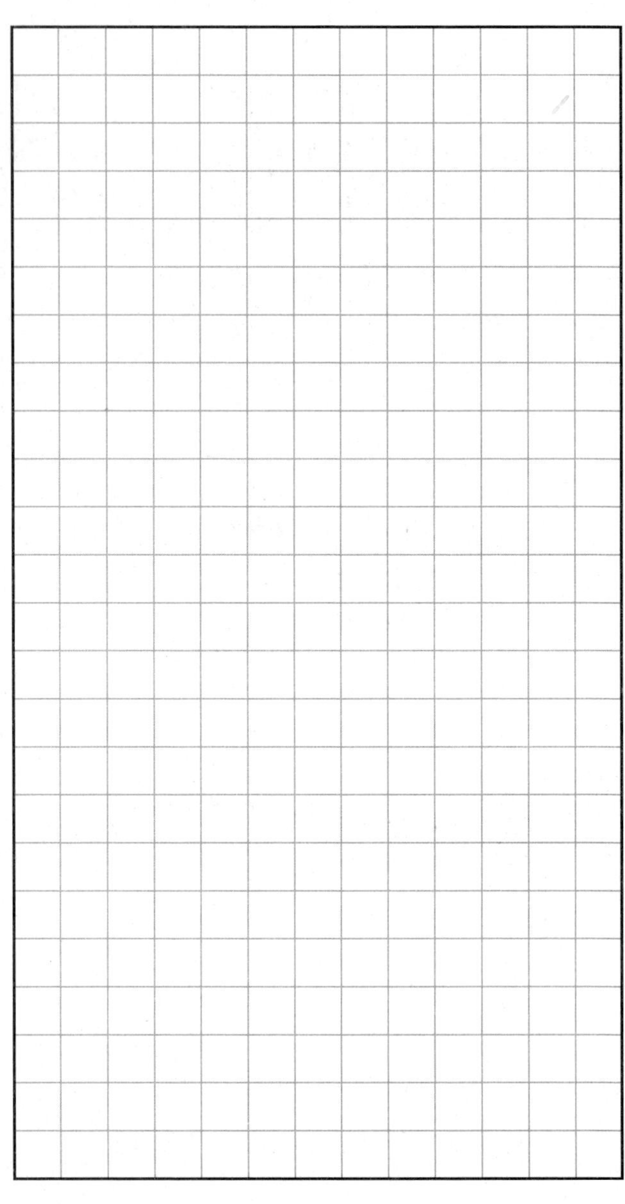

Name ____________________________________

Rough Sketch #3

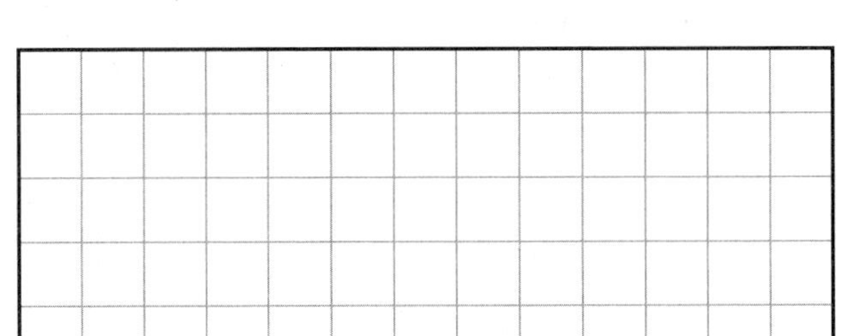

Refined Sketch

Discussion notes:

____

____

____

____

____

____

Comprehensive Sketch

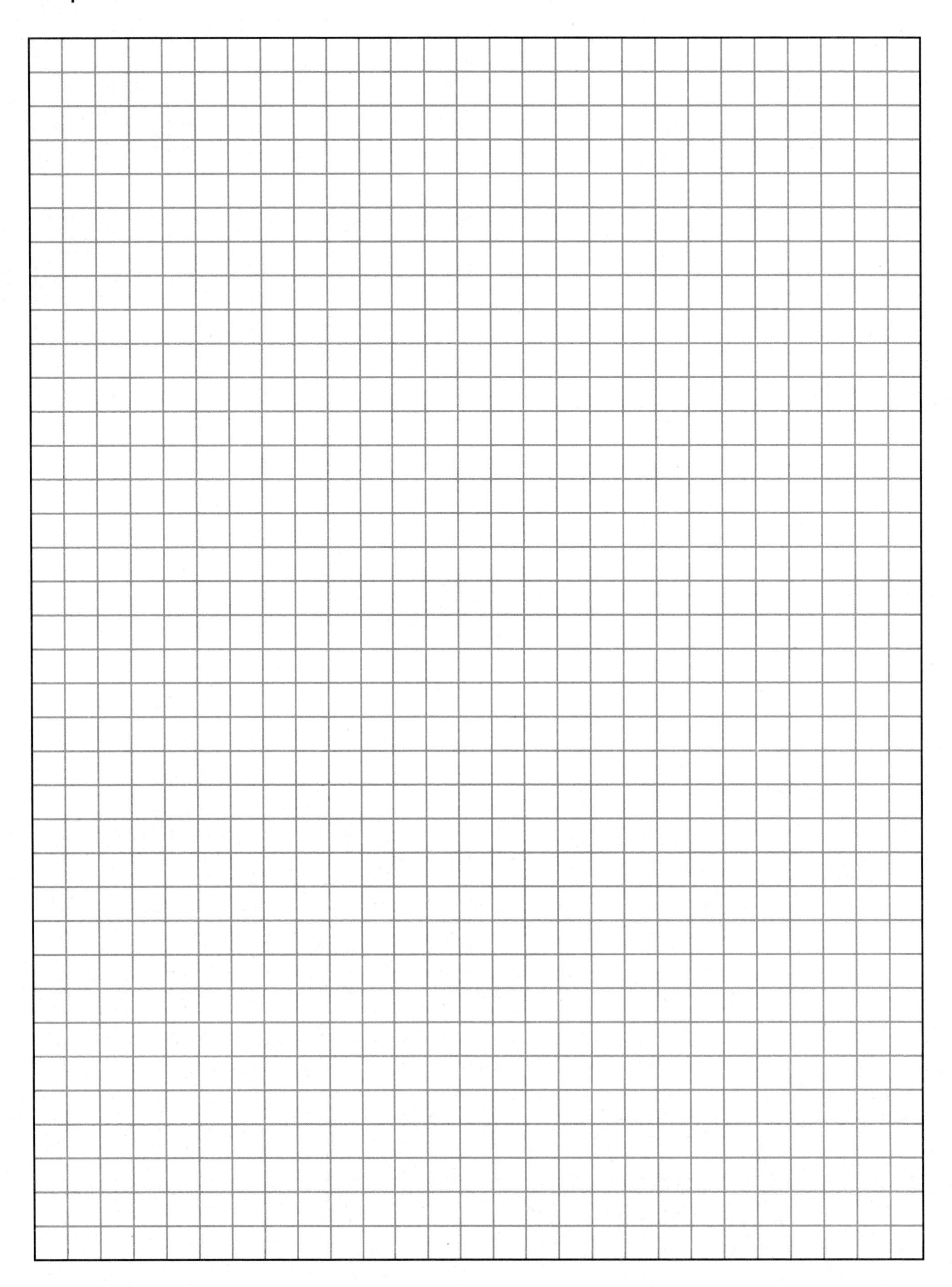

Name ________________________ Date __________ Class ________________

ENGINEERING DESIGN CHALLENGE

Prosthesis Challenge

Read the Engineering Design Challenge *presented at the end of Chapter 26 in the textbook before completing this worksheet.*

1. Identify the criteria and constraints which must be taken into consideration when you seek to respond to the problem statement. List the criteria and constraints on the following lines.

____
____
____
____
____
____

2. Write a problem statement for this design challenge using the information included and your answer to the previous question. Make sure to answer the who, what, where, when, and why.

____
____
____
____

3. Sketch two possible prosthetic limb solutions. Identify the available materials and ensure your solutions use only the provided materials. Present your ideas to classmates and collect their feedback and suggestions for improvement.

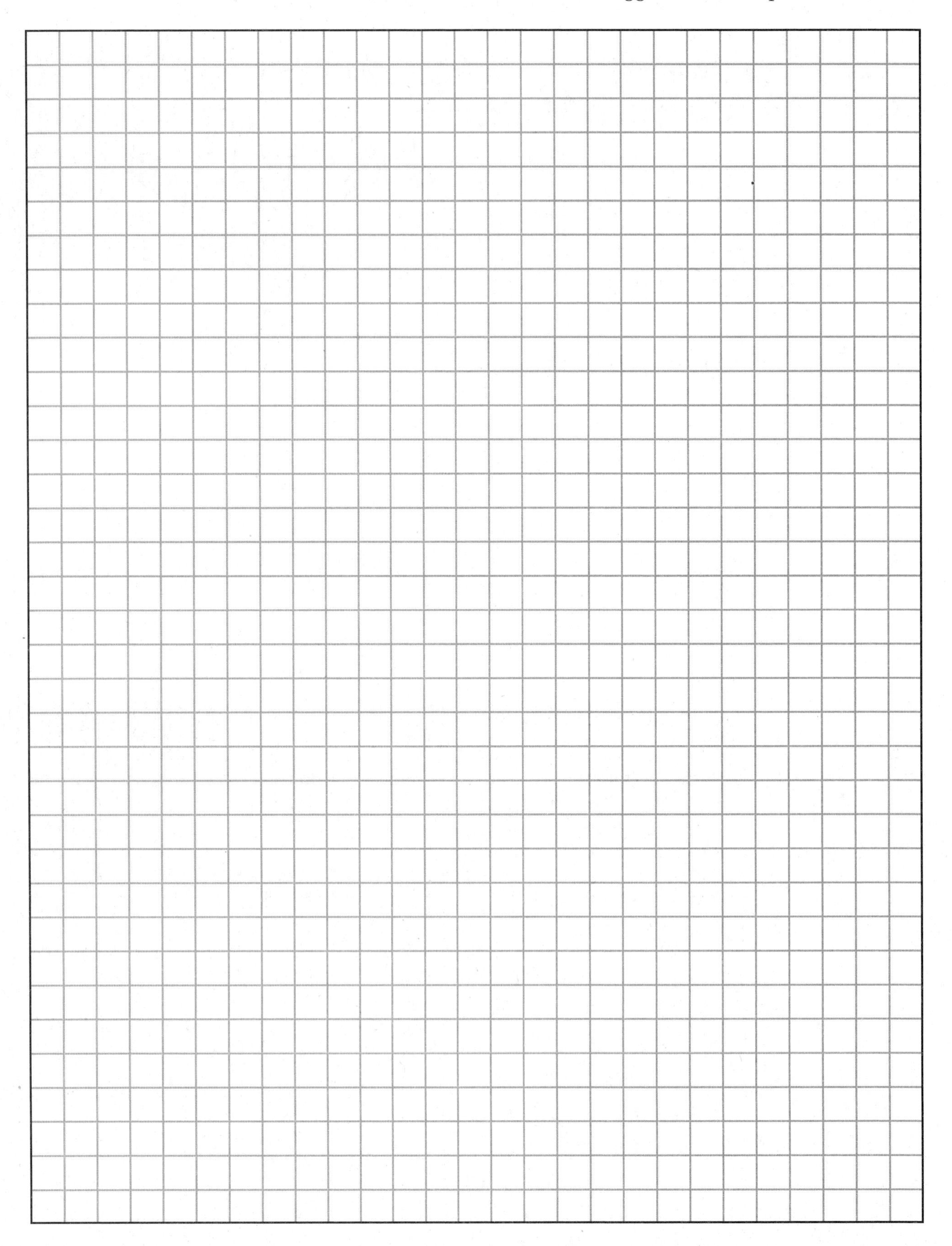

4. Prototype one of your solutions and test the prototype in performing the identified task (allowing a user with no arms to take a drink of water). Record the results of your testing.

____

____

____

____

____

____

5. Sketch your final solution. If you have access to design software, develop a digital model of your idea using the software.

Name ______________________ Date __________ Class ______________

ENGINEERING DESIGN CHALLENGE

Dehydrating Food

Read the Engineering Design Challenge *presented at the end of Chapter 27 in the textbook before completing this worksheet.*

Problem: ______________________

Fruit or vegetable selected to preserve by drying: ______________________

Amount of produce available: ______________________

Highest temperature of the dryer: ______________________

Drying time: ______________________

Amount of dried produce: ______________________

Have several people taste test the dried produce. Record their comments:

Tester # 1

Name: ______________________

Comments: ______________________

____

____

____

Tester # 2

Name: ______________________

Comments: ______________________

____

____

____

Tester # 3

Name:____

Comments: ____

____

____

____

____

Name ______________________ Date __________ Class ______________

ENGINEERING DESIGN CHALLENGE

Exothermic Reactions

Read the Engineering Design Challenge *presented at the end of Chapter 28 in the textbook before completing this worksheet.*

1. In addition to the *Embrace Warmer* described in the *Engineering Design Challenge,* what other problems or situations might need heat? Talk with others, conduct research, and brainstorm several ideas. Record your ideas on the following lines.

____

____

____

____

____

____

____

____

____

____

____

2. What types of reactants can be used to produce an exothermic reaction? List several reactants, the nature of the reaction, and the amount of heat produced (example on first line).

Reactant	Nature of the reaction	Heat produced
Iron powder, salt, water, carbon	Rust is created when exposed to air	Up to 163 degrees (Farenheit)

3. Identify one scenario (from the ideas you listed in the first questions or another idea) and one exothermic reaction. Sketch a device you could construct that would use the exothermic reaction to solve the identified problem.

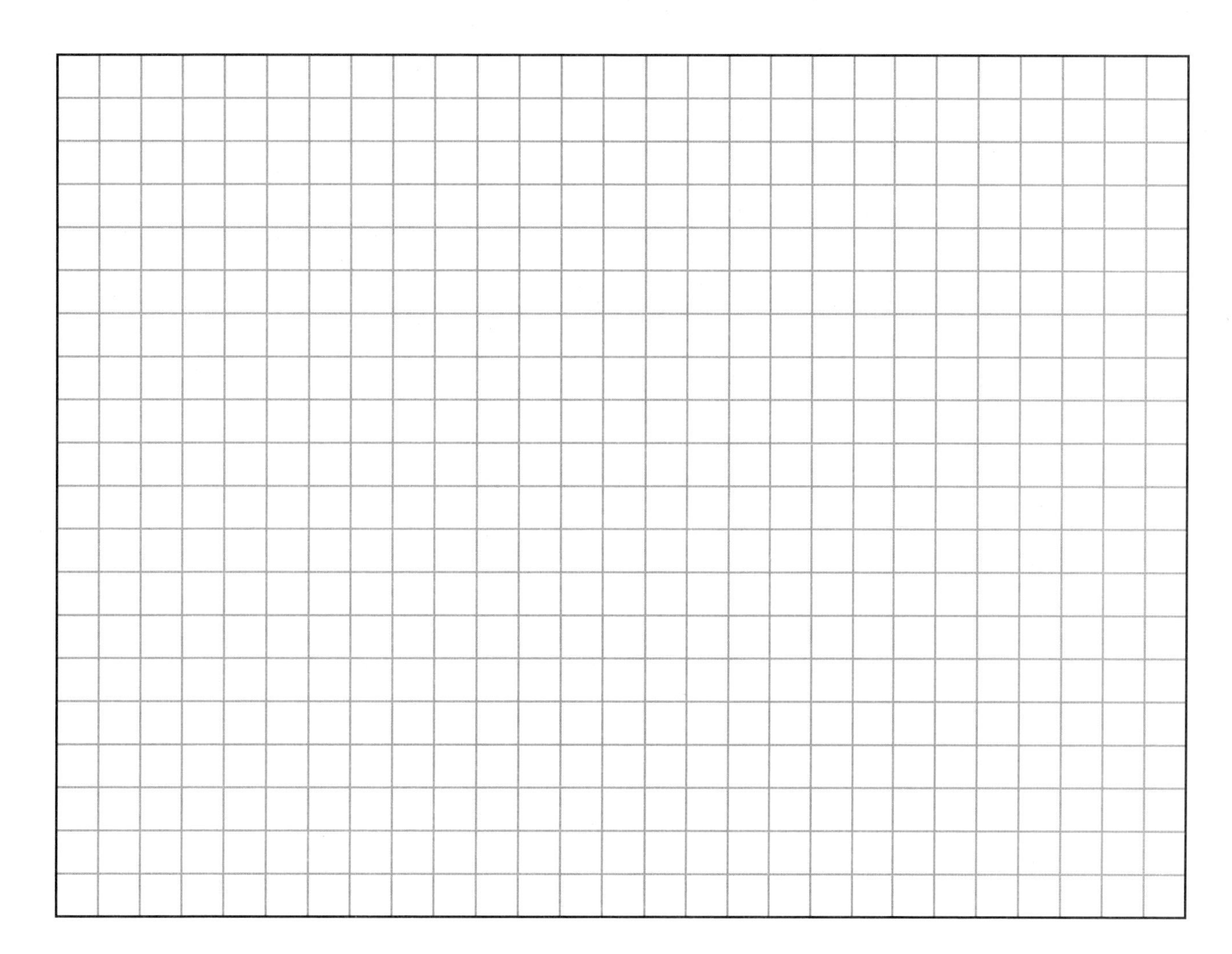

4. Describe how the device would work. How would the exothermic reaction happen? What is the desired outcome of your solution?

Name ________________________ Date __________ Class ________________

ENGINEERING DESIGN CHALLENGE

Resource Depletion

Read the Engineering Design Challenge *presented at the end of Chapter 29 in the textbook before completing this worksheet.*

1. Identify and define the problem. Then, write a problem statement. Make sure to answer the who, what, where, when, and why.

   ____

   ____

   ____

   ____

2. Research current methods of construction. Identify both environmentally friendly and harmful practices, approaches, and methods. Describe the best practices on the following lines.

   ____

   ____

   ____

   ____

   ____

   ____

3. Sketch a few possible home designs in the space provided. Identify the environmentally friendly features and potential problems of each design. Present your ideas to a classmate. Ask them to evaluate your designs using the Desired Outcomes included in the *Engineering Design Challenge* and to provide feedback on your ideas.

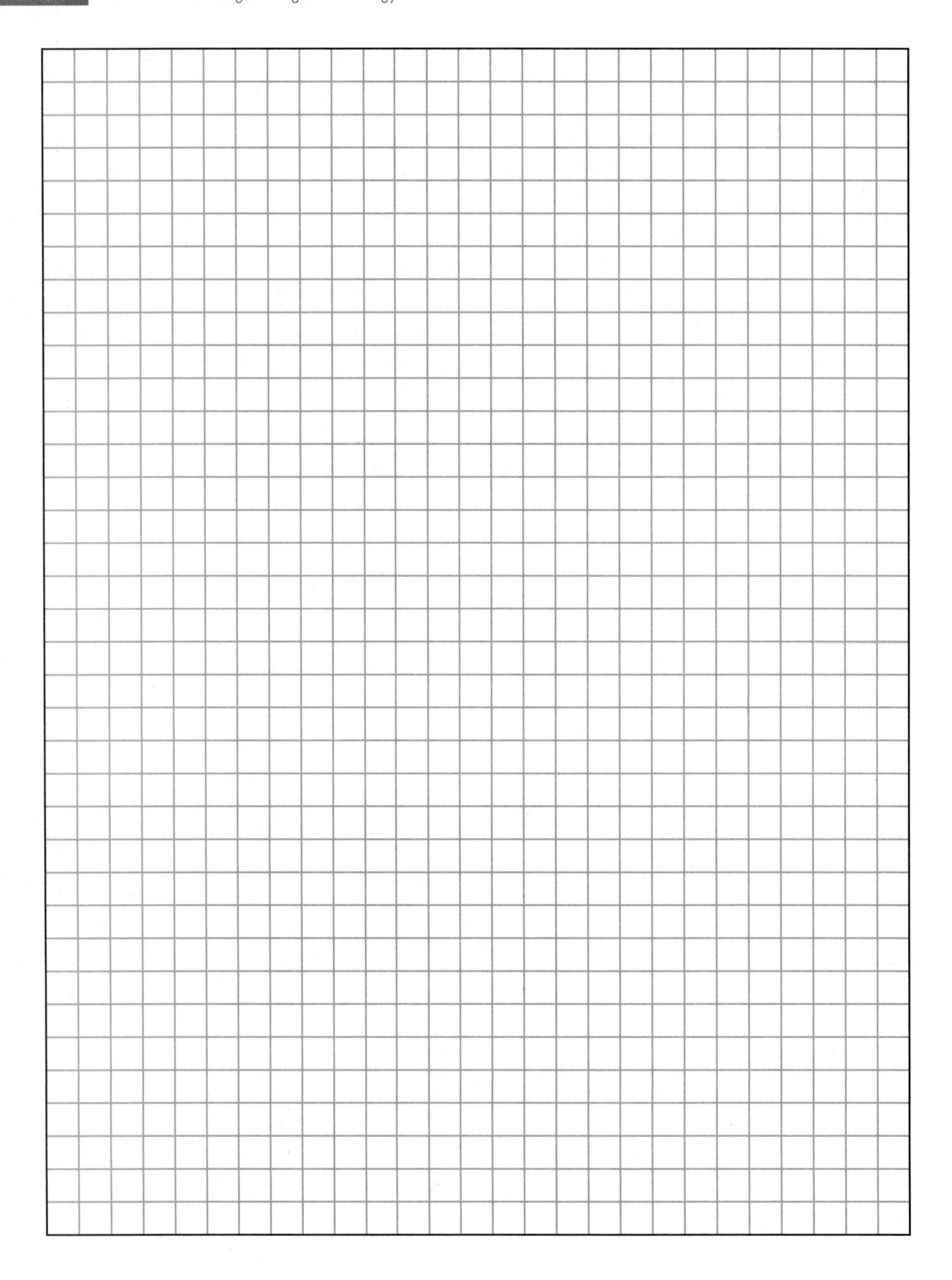

Name ______________________________

4. Research, identify, and record standards, regulations, and practices that could improve the environmental impact of new housing projects. Record these ideas on the following lines.

____

____

____

____

____

____

5. Using the environmentally friendly practices, approaches, and regulations you identified, the feedback provided, and the Desired Outcomes, determine the final design for your home. Sketch this design in the space provided. If you have access to a computer and design/modeling software, develop a digital model using the software.

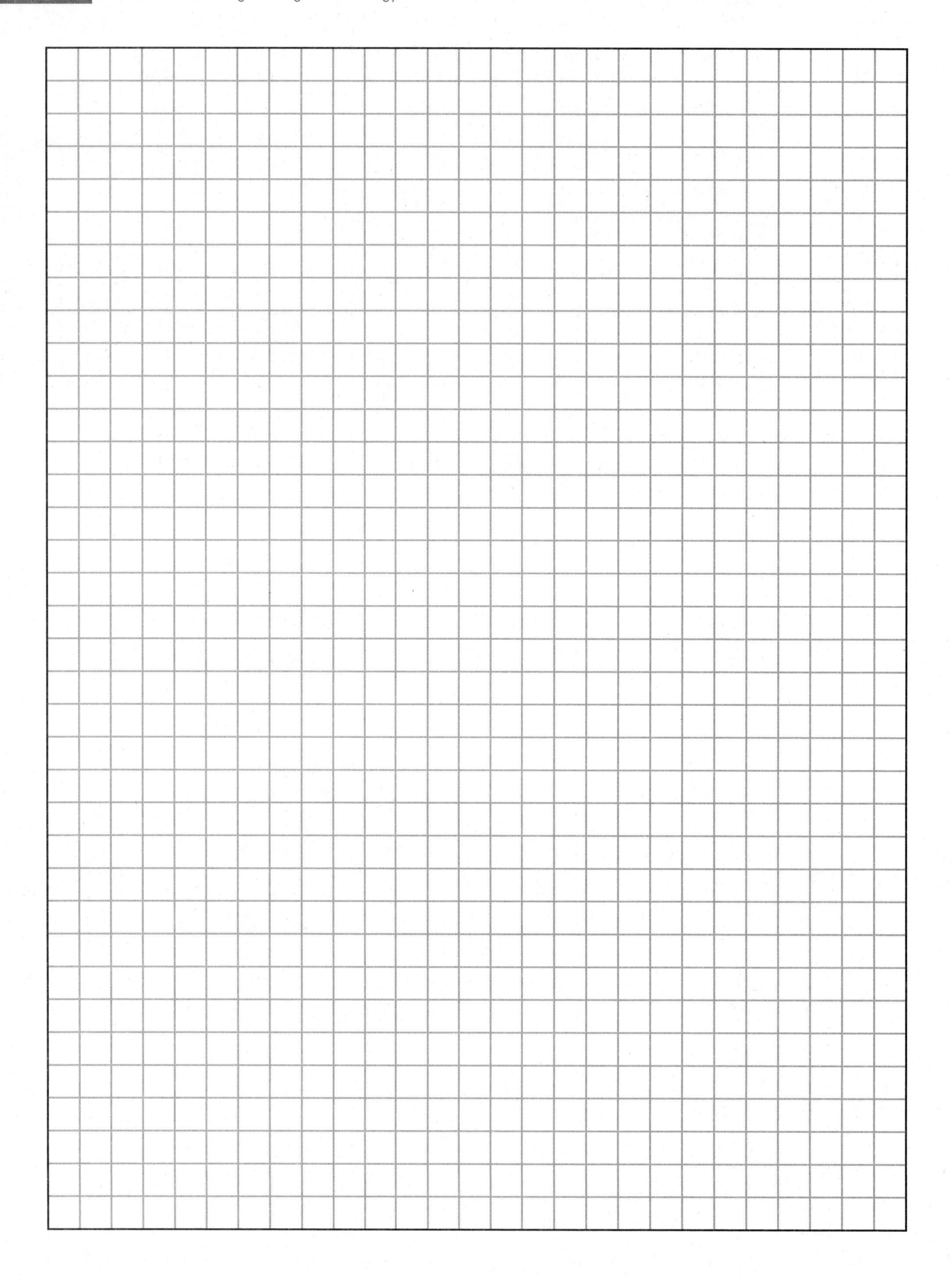

Name ______________________ Date __________ Class ______________

ENGINEERING DESIGN CHALLENGE

Technological Impacts Commercial

Read the Engineering Design Challenge *presented at the end of Chapter 30 in the textbook before completing this worksheet.*

Use the following area to develop a storyboard for a TV commercial that promotes recycling. Remember that a storyboard is a series of sketches showing all the scenes in a commercial. Dialog and production notes for each scene are included under each sketch. Use additional sheets to complete your storyboard, as necessary.

Series Title: ______________________

Description:

Description:

Description:

Description:

Description:

Description:

Name ____________________________________

Description:

Description:

Description:

Description:

Description:

Description:

Description:

Description:

Name ______________________________ Date ____________ Class ____________________

ENGINEERING DESIGN CHALLENGE

Forming a Company

Read the Engineering Design Challenge *presented at the end of Chapter 31 in the textbook before completing this worksheet.*

1. List the important tasks needed to design, produce, and sell a personalized calendar.

____

____

____

____

____

____

____

____

____

____

____

____

____

____

2. Draw an organization chart for a company that will design, produce, and market a personalized calendar. The chart should include all departments and show the chain of command, including levels of authority and responsibility.

Name ________________________ Date ____________ Class ________________

ENGINEERING DESIGN CHALLENGE

Operating a Company

Read the Engineering Design Challenge *presented at the end of Chapter 32 in the textbook before completing this worksheet.*

Developing the Format

Determine the parameters for the calendar using the form that follows.

Calendar Format

Paper size: ____________ width ____________ length

Page direction: ☐ Vertical ☐ Horizontal

Format: ☐ Pictorial ☐ Full page ☐ Two month

Day arrangement: ☐ Sunday–Saturday ☐ Monday–Sunday

Date rectangles: ☐ Square corners ☐ Rounded corners

Date location: ☐ Upper-left corner ☐ Upper-right corner ☐ Lower-left corner ☐ Lower-right corner

Text formats (fonts):

Text: ____________________

Credit lines: ____________________

Month: ____________________

Day of the week: ____________________

Date: ____________________

Credit lines: ____________________

Pattern in blank date boxes: ____________________

Other notes:

Promoting Products

On the following grid, develop a layout for a poster promoting your product.

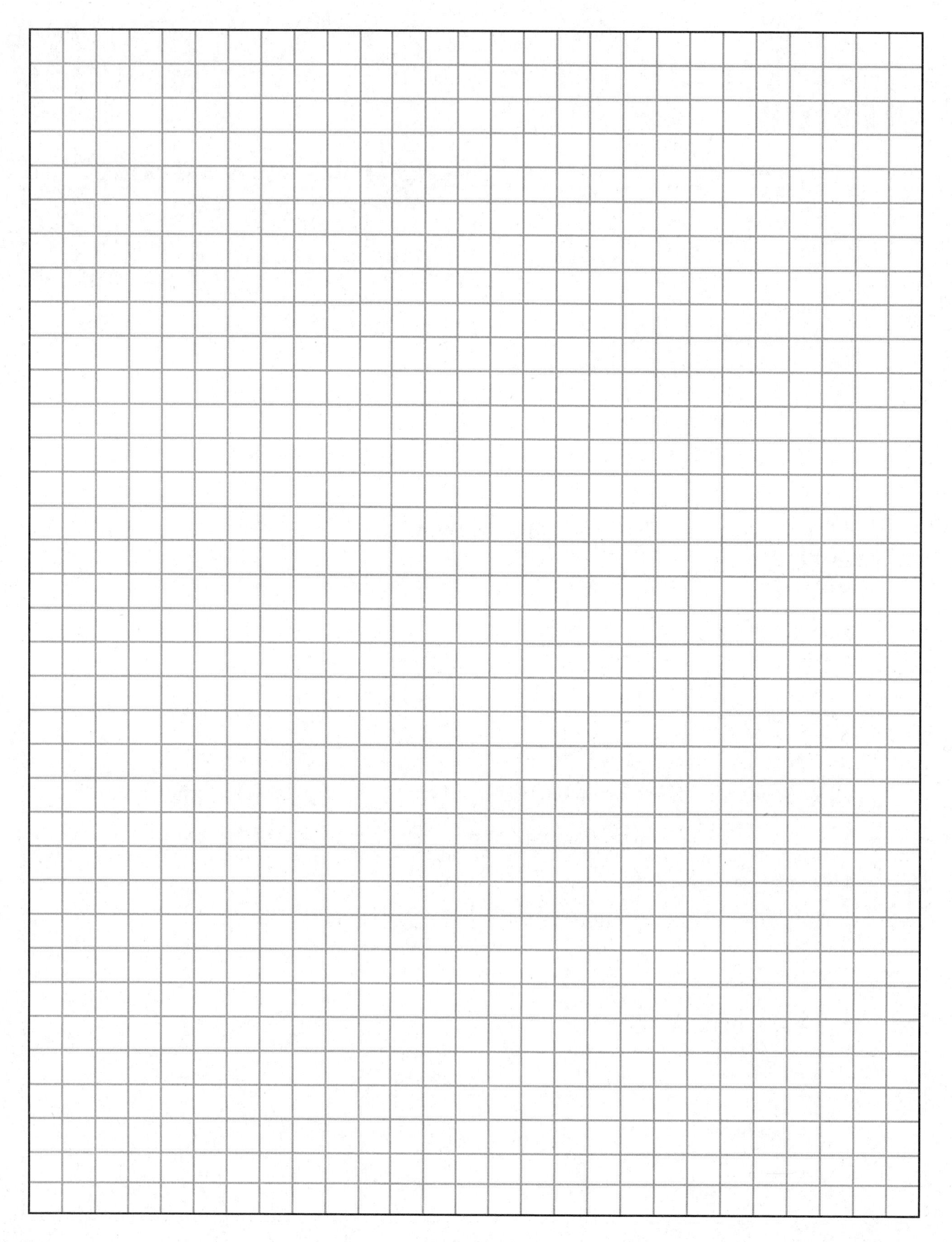

Name ___________________________________

Develop a form to sell entries on your personalized calendar. Alternatively, you can use the following forms.

Customer's name: ______________________

Address: ______________ City and state: ______________

Telephone number: (____) ______________

Calendar entry:

Date—Month: __________ Day: ________

Credit line entry: ______________

Customer's signature: ______________________

Cost per entry: **$** __________

Received: **$**

Customer's name: ______________________

Address: ______________ City and state: ______________

Telephone number: (____) ______________

Calendar entry:

Date—Month: __________ Day: ________

Credit line entry: ______________

Customer's signature: ______________________

Cost per entry: **$** __________

Received: **$**

Customer's name: ______________________

Address: ______________ City and state: ______________

Telephone number: (____) ______________

Calendar entry:

Date—Month: __________ Day: ________

Credit line entry: ______________

Customer's signature: ______________________

Cost per entry: **$** __________

Received: **$**

Name ____________________

Developing Financial Plans

Develop a budget (an estimate of income and expenses) for your company.

Income Budget

Number of products to sell . . . . ____________

Selling price . . . . $____________

Estimated income. . . . . $ ____________ (+)

Expense Budget

Production Costs

Image carriers (printing plates) . . . . $____________

Paper and other supplies . . . . $____________

Labor (hrs) x ($ per hour) . . . . $____________

Total production costs . . . . $ ____________ (−)

Marketing Costs

Advertising materials . . . . $____________

Sales commissions ($ per sale) x (sales) $____________

Total marketing costs . . . . $ ____________ (−)

Development Costs

Software purchase or rental . . . . $____________

Prototype (original model) material . . . . $____________

Additional costs . . . . $____________

Total development costs . . . . $ ____________ (−)

Total Costs . . . . $ ____________

Maintain a stockholder record using the following form.

Stockholder's Ledger			
Stockholder's Name	**Address**	**Date Purchased**	**Number of Shares**

Name ______________________________

Develop a form to sell your finished personalized calendar. Alternatively, you can use the following forms.

Customer's name: ______________________________

Address: ______________________ City or state: ______________________

Telephone number: (____) ____________

Number of Products	Cost per Product	Total Sale
	$	$

Salesperson ______________________________

Customer's name: ______________________________

Address: ______________________ City or state: ______________________

Telephone number: (____) ____________

Number of Products	Cost per Product	Total Sale
	$	$

Salesperson ______________________________

Customer's name: ______________________________

Address: ______________________ City or state: ______________________

Telephone number: (____) ____________

Number of Products	Cost per Product	Total Sale
	$	$

Salesperson ______________________________

Maintain a record of sales using the following form.

Sales Records											
Salesperson's Name	Sales by Day										Total
	1	2	3	4	5	6	7	8	9	10	

Name ________________________________

Purchasing Materials

Use the following form to inform the Purchasing Department of your material needs.

Purchase Requisition

Recommended vendor

Company's name: ________________________________

Address: ________________________________

City, state and zip: ________________________________

Material needed:

Quantity	Description	Estimated Cost

Authorization

Charge to budget code number: ______________

Approved by Department Head: ________________________________

Approved by Finance Vice President: ________________________________

Maintain a record of all financial transactions in a general ledger.

General Ledger					
Date	Entry Description	Income		Expense	

Name ______________________ Date __________ Class ______________

ENGINEERING & TECHNOLOGY ACTIVITY

Tension and Compression

Objective: Students will understand, experience, and be able to identify the difference between tension and compression, as well as the attributes of different materials under stress.

Introduction: Two forces acting on any building material are *tension* and *compression.* Each material has the capability of withstanding a certain amount of tension and compression. A tension force is one that pulls materials apart, while a compression force is one that pushes materials together.

Materials:

- 1′ long 1/4″ wooden dowel
- 1′ fishing line
- 2″ wooden cube
- Small stone/rock
- Paper clip

Safety: Always wear safety glasses when working with materials and testing. Wear the proper protective equipment and clothing when dealing with sharp objects.

Procedure: With your understanding of compression and tensional forces, form a hypothesis about each of the items in the Materials list. Predict whether each material will be *Strong under Compression* or *Strong under Tension* and place it in the corresponding column in the following table.

Strong under Compression	Strong under Tension

Testing: Test each material and record the results in the space provided.

Compress each material by squeezing it together. A material that is strong under compression will resist your efforts and change very little.

Next, test each material under tension by pulling the material apart. A material that is strong under tension will resist your efforts and change very little.

Material	Compression Testing Results	Tension Testing Results
Wooden dowel		
Fishing line		
Wooden cube		
Stone/rock		
Paper clip		

Reflection/Analysis Questions: Complete the following questions using your experience with testing the materials, tension, and compression.

1. Why might some materials be strong under compression while others are strong under tension?

    ____

    ____

    ____

    ____

2. What types of building scenarios would require a material that is strong under compression?

    ____

    ____

    ____

    ____

3. What types of building scenarios would require a material that is strong under tension?

    ____

    ____

    ____

    ____

4. Can you think of a material that is strong under both tension and compression?

    ____

    ____

    ____

Name ______________________________ Date __________ Class __________________

ENGINEERING & TECHNOLOGY ACTIVITY

Logic

Objective: Students will understand and be able to correctly produce logic statements involving the following conditionals: if, then, and or.

Introduction: Factories use automated control to govern their systems and processes. Part of system maintenance and setup requires an understanding of logic programming—a way of expressing things based on rules, relationships, and facts. One simple type of programming statement is called a *conditional statement*, which means there is a condition that will help determine the next action. For example, "If the sky is cloudy, then I will bring an umbrella." In this statement, the status of the sky is the condition that helps determine whether or not we bring an umbrella (the action). You can form these statements by using **IF _____, THEN _____**.

Materials:

- One deck of playing cards for a card game

Practice: Create a series of IF, THEN conditional statements you might use to determine what you choose to wear each day. For example, "If I have P.E., then I will wear my tennis shoes to school."

If ____,

then ____.

If ____,

then ____.

If ____,

then ____.

If ____,

then ____.

If ____,

then ____.

Next, work on combining your conditional statements to form more advanced rules and statements by adding AND/OR to the statements. For example, "If the sky is cloudy or I don't have a ride to school, then I will bring an umbrella," or "If the sky is cloudy and I have a field trip, then I will bring an umbrella." Record conditional statements related to choosing what you wear each day using IF, THEN, and AND/OR.

If ______________________________

and/or ______________________________,

then ______________________________.

If ______________________________

and/or ______________________________,

then ______________________________.

If ______________________________

and/or ______________________________,

then ______________________________.

If ______________________________

and/or ______________________________,

then ______________________________.

If ______________________________

and/or ______________________________,

then ______________________________.

Application: Using a deck of playing cards, teach a partner or friend how to play a card game. Practice playing the game until everyone is comfortable with the rules.

Using the space provided, write the rules to the card game using conditional statements. For example: *If you draw a card lower than your partner, then you lose your turn.* Use statements that are as simple or advanced as necessary to convey the rules of the game.

Once the rules are complete, try to teach someone else how to play the game using only the conditional statements you have written.

____

____

____

____

____

____

Name ______________________________ Date __________ Class ____________________

ENGINEERING & TECHNOLOGY ACTIVITY

Simple Machines

Objective: Students will be able to correctly identify and utilize each of the simple machines to form a complex machine.

Introduction: Simple machines are basic mechanisms which change the direction or magnitude of a force and make work easier for us to do.

Materials:

- 1 Tennis ball
- 1 Roll of masking tape
- 1 Broom
- 3′ Piece of string
- 3 Marbles
- 1 Ruler
- 5 Pieces of cardboard, each measuring 12″ × 12″
- 1 Piece of pine 2″ × 4″ that is 12″ in length
- 5 Eye bolt thread-eye screws (10 mm × hoop 3 mm)

Safety: Always wear safety glasses when working with materials and testing. Wear the proper protective equipment and clothing when dealing with sharp objects.

Practice: Simple machines make life easier by allowing us to multiply/change our effort to complete tasks more effectively. For each of the simple machines listed, identify one common example and describe how it makes work easier (example on first line).

Simple Machine	Everyday Example	How It Makes Work Easier
Wheel & Axle	*Car wheel*	*Allows us to move heavy things with less effort*
Inclined Plane		
Wedge		
Lever		
Pulley		
Screw		

Now, take your knowledge of simple machines and put it to work. You must use only materials provided to build a simple machine that will move a tennis ball from one location to another. Starting your simple machine must require nothing more than the work of one finger.

Name ___________________________________

Use the space provided to take notes, draw potential solutions, and record results.

Draw your final simple machine in the sketch box below.

Name ________________________________ Date ___________ Class ____________________

Objective: Students will be able to describe the principle of buoyancy and predict whether objects will sink or float based on their understanding.

Introduction: An object's buoyancy describes how well it floats in a fluid. If an object floats in water, the upward buoyant force of the water is greater than the downward force of the weight of the object. Submarines use ballast tanks filled with air or water to regulate their weight in order to rise, sink, or remain neutrally buoyant.

Materials:

- Ruler
- Frisbee
- Pen
- Can of diet carbonated soda
- Can of regular carbonated soda
- 20 Golf balls
- 1 Piece of tin foil measuring 12″ × 12″
- 12″ Length of masking tape
- Large container (at least 12″ × 18″ × 6″) full of water

Will It Float? Objects that appear very large may float easily (think of a ship), while objects that are much smaller sink (marbles). An object's ability to float or sink depends not only on its weight, but also on the amount of water it displaces and its density.

Make a prediction for each object listed in the following table—will it sink or float?

After noting your guesses, place each object in the container of water. Record whether it sinks or floats in the *Test Result column of the table.*

Object	Sink?	Float?	Test Result
Ruler			
Frisbee			
Pen			
Can of diet carbonated soda			
Can of regular carbonated soda			
Golf ball			

Activity: An object's ability to float can be increased by reducing its density. One approach to reduce density involves increasing the surface area of an object. For example, a flat piece of metal will float, but a solid metal brick may sink.

Using the materials listed at the beginning of this activity, construct a "boat" that can hold as many golf balls as possible without sinking. Before building the boat, determine the shape that will allow you to maximize the surface area. Use the space provided to record ideas and sketch the shape and surface area of your boat.

Test your boat and record the results.

Boat shape (draw a picture):

Boat dimensions:

Length: ______________

Width: ______________

Height: ______________

Number of golf balls held: ______________

What changes would you make next time?

____

____

____

____

Name ______________________ Date ____________ Class ______________

ENGINEERING & TECHNOLOGY ACTIVITY

Paper Towers

Objective: Students will explore the qualities required to construct structures, including shape and material.

Introduction: Humans have long been fascinated with building structures that are stronger and taller than any before. Take a look at the skyline of any major city and you will see a variety of building approaches. Designing tall buildings that will withstand the elements can be very challenging.

Materials:

- 10 Sheets of 8.5″ × 11″ paper
- Roll of masking tape
- Scissors
- 6 Weights that are 1/2 pound each

Safety: Always wear safety glasses when working with materials and testing. Wear the proper protective equipment and clothing when dealing with sharp objects and scissors.

Testing: Your objective is to build a tower, using 10 pieces of paper and tape. The tower must be at least 18″ tall and must hold at least 3 lbs. In order for your tower to succeed, it must be both stable and strong.

What qualities or approaches would improve the stability of your tower?

____

____

____

What qualities or approaches to building would improve the strength of your tower?

____

____

____

What inspiration can you take from towers, or other tall structures, you have seen in person? How could the same building approaches be applied to your tower?

____

____

____

Sketch several possible shapes and ideas for your tower building in the space provided.

Before you begin building, experiment with a variety of designs and approaches. Try building each design and test how many weights can be placed on top before the tower fails. Record your trial results.

Trial #1

Sketch of your tower:

What went well?

____

____

____

____

____

What didn't work?

____

____

____

____

____

Trial #2

Sketch of your tower:

What went well?

____

____

____

____

____

What didn't work?

____

____

____

____

____

Trial #3

Sketch of your tower:

What went well?

____

____

____

____

____

What didn't work?

____

____

____

____

____

Trial #4

Sketch of your tower:

What went well?

____

____

____

____

____

What didn't work?

____

____

____

____

____

Using your trial results, make decisions about which approach may be the most successful. Decide on your final design and create a sketch of your tower. Test your tower and record the results.

Final tower design:

How much weight did the tower hold?

Why was this design successful?

Name ______________________ Date __________ Class ______________

ENGINEERING & TECHNOLOGY ACTIVITY

Power Plants

Objective: Students will research and identify the strengths and weaknesses of several naturally-occurring resources. Students will research several types of power plants and produce a power plant diagram.

Introduction: The conversion of raw materials (e.g., coal, oil, etc.) into usable power and electricity is critical to ensure continuity and comfort in daily living. Power plants are specialized facilities that harness energy from elements and convert it into electricity for transmission to customers. Each power plant is uniquely designed to harness a particular form of energy and convert it into usable energy. For example, a power plant located on a river may be designed to harness the energy from falling water to turn a wheel and generator, while a wind generator must be positioned along the proper wind currents to turn the turbines.

Materials:

- One large piece of paper (at least 12″ × 12″)
- Colored pencils
- Ruler

Identification: Several naturally-occurring resources or phenomena have been harnessed or collected for power production. In the following table, list five naturally-occurring resources or phenomena that can be used to produce power.

Resource/ Phenomena	Where is it harnessed/ collected?	How can it be used to produce power?

Conversion: Most power plants use a generator to convert mechanical motion (something moving) into usable power. For example, a stream may turn a waterwheel, which rotates a shaft connected to a generator. As the shaft is turned, the generator forces the movement of electrical charges, which produces electricity through the electromagnetic induction.

Some forms of energy produce higher amounts of electricity using more efficient processes with less by-product (pollution, waste, etc.). Using the Internet, library, or a textbook, identify the strengths and weaknesses of each of the resources you listed in the previous section.

Resource/Phenomena	Strengths	Weaknesses

Name ____________________________________

Practice. Choose one resource/phenomena from your list. In the space provided, draw a diagram of a power plant that shows the process of converting the resource/phenomena from its natural state into usable energy.

Name ______________________ Date __________ Class ______________

ENGINEERING & TECHNOLOGY ACTIVITY

Transportation Systems

Objective: Students will research transportation systems and design their own transportation system for a specified location.

Introduction: As a growing number of individuals work in centralized cities and buildings, the development of sophisticated transportation systems has become increasingly important. Trains, airplanes, buses, and automobiles must all work together in a larger transportation system to ensure effective and safe transportation for all. A glitch or problem in a transportation system can cause widespread delays, problems, and even danger for many people.

Materials:

- One large piece of paper (at least 12″ × 12″)
- Colored pencils
- Ruler

Identification: Each form of transportation has unique advantages and disadvantages. Identify the advantages and disadvantages of each form of transportation listed in the following table. Consider factors such as speed, convenience, flexibility, demand, and any others you may think of.

Transportation System	Advantages	Disadvantages
Bus		
Train		
Airplane		
Automobile		
Boat		
Other		

In addition to vehicles, there are many other items that make up a transportation system as a whole. Items such as signs, markings, symbols, schedules, and rules all work toward an effective system that allows users to move safely. List several non-vehicle components of transportation systems in the following table. Draw a picture of each component and describe its function.

Item	Illustration	Function

Design: Imagine that you have been asked to develop a new transportation system for a city that will be built on a previously uninhabited island. Think about the vehicles, areas of travel, signs, symbols, marking, rules, and procedures you will put in place to keep everyone safe. Sketch your proposed solution on the following map of the island. Make notes about the related rules, procedures, and other items that will contribute to the functioning of the transportation system.

Name ______________________________ Date ___________ Class ____________________

ENGINEERING & TECHNOLOGY ACTIVITY

Water Refraction

Objective: Students will experiment with and experience light refraction. Students will produce an infomercial about light refraction.

Introduction: Have you ever noticed that a straw in a glass of water appears to be bent or broken? This is due to the effect of light refracting as it moves through one material and into the next.

Materials:

- One clear glass jar
- Water
- Pencil
- Printed photograph or drawing of a simple object (basic shape, arrow, emoji, etc.)
- Scissors

Safety: Always wear safety glasses when working with materials and testing. Wear the proper protective equipment and clothing when dealing with sharp objects.

Refraction. When light travels through a given medium, it travels in a straight line. When light passes from one medium to another, however, the path the light takes bends (refracts). This bend occurs at the boundary where light travels from one medium to another. Once the light moves to the new medium, it travels in a straight line again.

Fill the glass jar halfway full with water and place the pencil into the jar until it touches the bottom. What do you notice? What happened? Test the pencil at different locations (center, side) and angles (straight down, angled). Record your observations.

____

____

____

____

____

____

Place the photograph or drawing behind the glass jar and move so that your eyes are level with the water level in the jar. What happens when you move the image up and down behind the jar? Why does this happen? Move the image closer to the jar and further away. What happens? Record your observations.

____

____

____

____

____

____

Angle. Another important element of light refraction relates to the viewer's angle. Try positioning yourself at different angles (i.e., above the jar, at eye level with the water, at a 45-degree angle, etc.) to view the pencil as it passes into the water. What do you notice? What happens as you move? Record your observations.

____

____

____

____

____

____

Suppose that you were hired to teach a group of novices how to spear fish. Spear fishing requires a fisherman to throw a spear at a fish. Understanding light refraction is key to successful spear fishing. In this case, the novice fishermen could not spear any fish, despite their best efforts. You decide to create a 30-second video explaining the concept of light refraction to share with them. The fishermen will watch the video on their phones before their next fishing lesson. Use the space provided to storyboard your video.

Character/Event	Dialog/Directions	Production Notes

Character/Event	Dialog/Directions	Production Notes

Name ______________________ Date __________ Class ______________

ENGINEERING & TECHNOLOGY ACTIVITY

Design for Accessibility

Objective: Students will learn about various disabilities and ways of designing everyday objects in an accessible manner.

Introduction: Do you know anyone who has a disability? Have you ever considered that there may be many things they cannot do/use because of the design? User-centered design is design focused on the end user of the product and emphasizes accessibility and ease of use. You will take several everyday-objects and redesign them with an emphasis on accessibility.

Materials:

- Computer mouse
- Computer keyboard
- Spoon
- Plastic drinking cup
- Pencil
- Light switch
- Book

Disabilities. A wide variety of disabilities exist, each with unique characteristics and effects. Pick one disability from the following list and circle it.

- Spina bifida
- Cerebral palsy
- Epilepsy
- Cystic fibrosis
- Multiple sclerosis
- Muscular dystrophy

Use the Internet to research the disability you chose. Answer the following questions about what you learned.

1. What are some of the ways this disability manifests?

2. What are some of the challenges associated with this disability? What activities or functions might be difficult to perform?

3. How would this disability change your personal daily routine?

User-Centered Design. Pick three of the objects from the Materials list. Brainstorm, test, design, evaluate, and redesign the objects with a user-centered approach related to the disability you chose. Document the process for each object you select.

Object #1: ____________________

How will you redesign it?

____

____

____

____

____

____

Record your notes, sketches, and observations in the space provided.

Object #2: ____

How will you redesign it?

____

____

____

____

____

____

Record your notes, sketches, and observations in the space provided.

Object #3: ____

How will you redesign it?

____

____

____

____

____

____

Record your notes, sketches, and observations in the space provided.

Name ______________________ Date ____________ Class ______________

ENGINEERING & TECHNOLOGY ACTIVITY

Biomimicry

Objective: Students will study, identify, and practice using biomimicry principles to complete an engineering design task.

Introduction: *Biomimicry* is a term that means we mimic life/nature. Engineers can use attributes, designs, and approaches from nature while designing. You will identify several traits in nature and then use biomimicry principles to complete a design challenge.

Materials:

- Nails
- Screws
- Glue
- Plywood
- Cardboard
- 1/8″ plastic
- Scissors
- Tape

Safety: Always wear safety glasses when working with materials and testing. Wear the proper protective equipment and clothing when dealing with sharp objects.

Biomimicry. Animals, plants, and structures found in nature can be valuable tools for inspiration around difficult problems.

Use the Internet to identify several challenges for which plants or animals have developed traits to overcome. Record your findings in the following table (example in first row).

Animal/Plant	Challenge	Solution
Moth	*It can be very difficult for a moth to see while flying at night. Additionally, a moth must be careful to not attract attention of predators.*	*The eyes of moths are covered in anti-reflective nanostructures. These prevent light from reflecting off the eyes when moths fly at night, which would give away their location. This adaptation helps them avoid predators.*

Design Challenge: Use biomimicry to work toward a solution for the following design problem. Many deaths each year can be attributed to accidents that occur while rock climbing and hiking. These deaths are often caused when climbers slip, fall, or lose their footing. Himalayan mountain goats may offer inspiration for the design of new gear that may help avoid accidents experienced by mountain climbers and hikers.

Use the space provided to record research notes, sketch, and design gear for climbers using biomimicry. When you have finished researching and designing, use the items in the Materials list to create a mock-up of your design.

Name ___________________________

Name ______________________ Date __________ Class ______________

ENGINEERING & TECHNOLOGY ACTIVITY

Architectural Trends

Objective: Students will develop an understanding of architectural trends and changes that have taken place over time. Students will use their understanding to create a time line that represents architectural trends and shifts.

eFesenko/Shutterstock.com

Introduction: Think of the homes in your neighborhood. Compare them to homes in other neighborhoods, other cities, and other countries. A variety of architectural styles and approaches have emerged over the years, from the stone columns and arches in ancient Rome and Greece to lines and angles created with glass, steel, and concrete in Postmodern architecture.

Materials:

- Large graph or unlined paper
- Colored pencils
- Ruler
- Scissors
- Markers
- Pencil

Safety: Proper care should be taken when working with sharp objects and scissors.

Architecture. Like fashion, architectural styles, designs, and trends come, go, and change over time. Research the common architectural styles listed in the following table, and identify key characteristics of each. Identify buildings in your community or beyond that are examples of each architectural style (example in first row).

Style	Defining Traits	Example
Georgian/Neoclassical	• *Simple geometric forms* • *Use of columns* • *Blank (often white) walls*	*United States White House*
Spanish Colonial		
Ranch		
Saltbox		
Art Deco		

Time Line Activity: Design and create an architectural trend time line using the items in the Materials list. The time line should include examples (text or image) of each style and be designed in such a way that it could be used as a learning tool. Use your completed table and the following list of architectural styles for reference. Sketch out ideas in the space provided before creating your final time line.

Name ______________________

- Saltbox
- Cape Cod
- Tidewater
- Spanish Colonial
- Georgian
- Federal/Adam
- Greek Revival
- Second Empire
- Stick
- Queen Anne
- Colonial Revival
- Spanish Eclectic
- Tudor Revival
- Foursquare
- Prairie
- Art Deco
- International
- Ranch
- Postmodern

Why do you think architectural trends have changed so much over time?

While making your time line, did you notice any trends that emerged in different time periods? Why might some trends come and go while others remain constant?

Which architectural styles or trends are your favorite? Why?

Name ______________________ Date __________ Class ______________

ENGINEERING & TECHNOLOGY ACTIVITY

Plastics from Milk

Objective: Students will learn about the properties of different materials, such as milk and plastic. Students will explore various procedures used in materials processing to turn milk into plastic.

Introduction: Choose a piece of plastic or plastic item and inspect it. How do you think it is made? Is all plastic the same? Did you know you can turn common milk into plastic?

Materials:

- 1 cup milk
- Stovetop and pan or microwave and microwavable container
- 4 teaspoons of white vinegar
- 6 paper towels
- 1 spoon
- Mold for plastic (optional)

Safety: Proper care should be taken when working with heat, flames, and other items that may cause burns. Safety glasses should be worn when working with chemicals.

Plastic. Plastics are found everywhere. They can be different sizes, shapes, textures, and strengths, but all plastics are made of polymers (repeating molecules in chains). Identify several different items made of plastic and the qualities of each in the following table.

Plastic Item	Physical Appearance (color, transparency, texture, etc.)	Ductility (ability to be stretched or compressed)	Resistance to Breaking

Milk to Plastic: Follow the directions provided to turn common drinking milk into plastic. The process involves adding an acid (vinegar) to warmed milk, which changes the pH of the milk. This change causes the casein molecules to break apart and reform into polymer chains. These chains are casein plastic.

1. Heat 1 cup of milk in a pan or the microwave until it is just barely boiling.
2. Add 4 teaspoons of vinegar.
3. Stir the mixture slowly.
4. Record your observations. What is happening?

____

____

____

____

____

5. Let the mixture cool for 5–10 minutes. Prepare a stack of three paper towels.
6. Use a spoon to scoop the curds out and onto the paper towels. Leave the liquid in the pan.
7. Use the remaining paper towels to press down on the curds and remove as much liquid as possible.
8. The curds have become casein plastic. Mold it to the desired shapes. Optionally use the molds to shape the plastic.
9. Allow to set for 48 hours.

Reflection: Why do you think this combination of materials produced plastic?

How are other plastics formed? Use the Internet to research plastic forming processes. Are there any common trends across the processes?

Name ______________________ Date __________ Class ______________

ENGINEERING & TECHNOLOGY ACTIVITY

Tumblewing Glider

Objective: In this activity, you will create a paper glider called a tumblewing, which demonstrates lift as it "tumbles," or rotates, through the air.

Introduction: Tumblewings work on the principle of ridge lift, which is exhibited as air is deflected upward off a surface. In this activity, you will create an area of ridge lift for your tumblewing to glide on.

Materials:

- Scissors
- One 2″–3″ piece of cellophane tape
- Tissue paper, newspaper, or other lightweight paper
- Book or binder

Safety: Always handle scissors carefully.

Procedure:

1. Gather items from the Materials list.
2. Cut out the tumblewing pattern provided and tape it to a piece of lightweight paper.
3. Trace firmly over the dotted lines of the pattern to indent or crease the lightweight paper. This will make the paper easier to fold.
4. Trace around the solid lines and remove the pattern.
5. Use scissors to cut out the tumblewing from the piece of paper. Set the pattern aside.
6. Gently fold the paper along the creases on each of the short ends, in the same direction.
7. Fold the edge of one long side up, and fold the edge of the other long side down.

8. Hold the tumblewing up in the air with the edge that is folded up facing toward you. Gently let go of the tumblewing and watch it rotate while it falls.
9. Hold a book or binder in your hands tilted at a slight upward angle. Walk forward. Feel the air as it is forced upward toward you.
10. While still holding the book or binder in one hand, hold the tumblewing in the other hand above your head. Drop the tumblewing in front of you while walking forward. This movement should provide the necessary ridge lift for the tumblewing to continue flight.

Note: If you walk too fast or slow, or tilt the book or binder at the wrong angle, the ridge lift is removed and the tumblewing will fall.

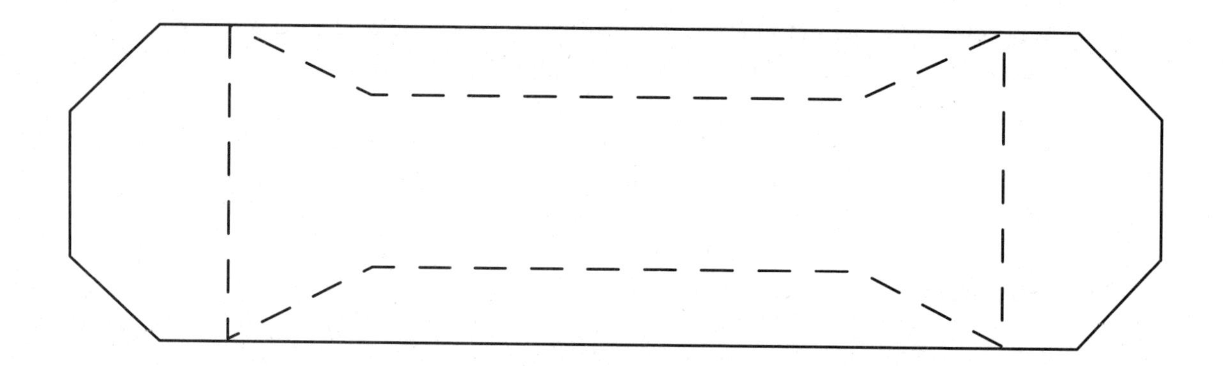

Name ______________________ Date __________ Class ______________

ENGINEERING & TECHNOLOGY ACTIVITY

Using and Assessing Products

Objective: Students will learn where to find information regarding the use of a product and explore the positive and negative impacts of that product.

Introduction: Many products come with owners' manuals that explain how to use a product. Some products might not need a manual, while others may require a more complex manual in order to help people use the product correctly. Work with two or three classmates to complete this activity.

Step 1: You will be assigned a product by your instructor. What information do you think you will need in order to use the product effectively? Consider the complexity of the product.

____

____

____

____

____

____

Step 2: Nearly all products impact people, society, and the environment in some way. These impacts can be positive or negative. List some impacts of your product.

Impacts on People

- Positive

____

____

____

____

- Negative

Impacts on Society

- Positive

- Negative

Impacts on the Environment

- Positive

- Negative

Name ________________________________ Date __________ Class ____________________

ENGINEERING & TECHNOLOGY ACTIVITY

Product Servicing

Objective: Students will learn how to service and repair a flashlight.

Introduction: You will be given a flashlight that does not light. The cause of the failure is likely to be in one or more of three areas:

- The batteries might be dead.
- The switch might not be making contact.
- The bulb might be burned out.

Materials:

- Flashlight
- Schematic of flashlight
- Volt-ohmmeter for testing
- Service report (provided)

Safety: Exercise the general safety rules used when working with electrical and electronic devices.

Balsum #601 Flashlight	**Service Manual**

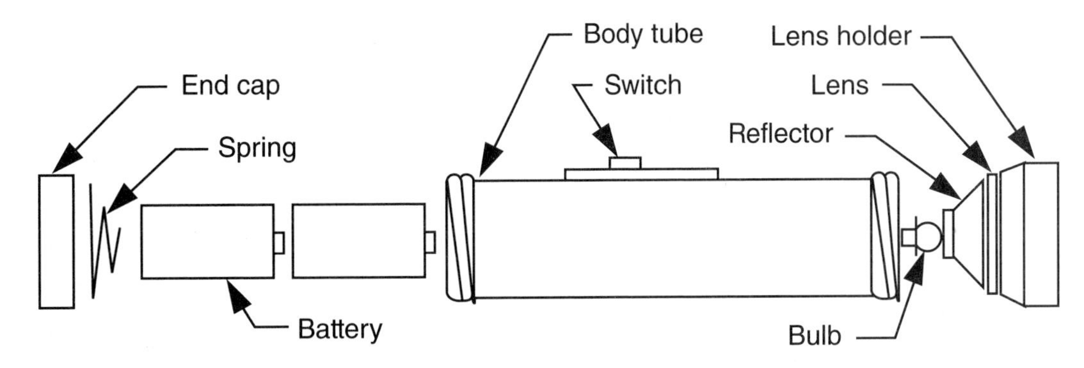

Parts Description

The Balsum #601 Flashlight seldom needs servicing. When it does, however, use the following procedure to complete troubleshooting and repair activities. This procedure follows a logical set of actions, which test the possible areas of failure in order of their probability, from most probable to least probable.

Service the product by performing the following:

1. Read the maintenance manual.
2. Service the product.
3. Complete the service report.

Testing the Batteries

1. Disassemble the flashlight and remove the batteries.
2. Set the volt-ohmmeter (VOM) at the 0–10 V direct current (DC) scale.
3. Place the (–) lead of the VOM on the bottom of the battery.
4. Place the (+) lead on the raised tip at the top of the battery.
5. A good battery produces a reading of 1.5 V on the VOM. If the batteries are bad, replace them.
6. Reassemble the flashlight and test its operation.
7. Complete the service report.

Testing the Bulb

1. Disassemble the flashlight and remove the bulb.
2. Set the VOM to check resistance.
3. Place the (–) lead of the VOM on the contact at the bottom of the bulb.
4. Place the (+) lead of the VOM on the metal side of the bulb.
5. If the VOM reads a maximum (infinite) resistance, the bulb is burned out. Replace the bulb if it is burned out.
6. Reassemble the flashlight and test its operation.
7. Complete the service report.

Testing the Switch

1. Disassemble the flashlight.
2. Set the VOM to check resistance.
3. Place the (–) lead of the VOM on one switch contact.
4. Place the (+) lead of the VOM on the other contact.
5. Move the switch control. At one setting, the VOM should read almost no resistance. At the other setting, the VOM should read a maximum resistance.

6. If the switch reads a maximum resistance at both settings, clean the contacts of the switch or replace the switch.
7. Reassemble the flashlight and test its operation.
8. Complete the service report.

Balsum Manufacturing — Service Report

Customer's name: ______________________________

Product name: ______________________ Model number: ______________

Date received: ____________

What's wrong with the product?

Test Results

TEST	OBSERVATIONS	CORRECTIVE ACTION TAKEN
Testing the battery		
Testing the bulb		
Testing the switch		

Customer Billing

Time start: ________ Time finished: ____________ Time worked: ____________

x (labor rate) ____________

Labor cost → [ ]

Parts Used

Item	Number	Cost	Total
Batteries			
Bulbs			
		Parts cost	

→ [ ]

Total Bill (parts + labor) [ ]

Name ______________________ Date __________ Class ______________

ENGINEERING & TECHNOLOGY ACTIVITY

Telecommunication

Objective: Students will create a script for a 30-second television advertisement. Your instructor will assign a topic for your advertisement.

Introduction: Recall that the goals of communication—to inform or educate, to persuade, to entertain, and/or to control or manage certain actions. Keep these goals in mind as you develop your advertisement.

Materials:

- Blank script form provided
- Pen or pencil

List your topic:

____

Develop three ideas for your topic. Remember, you have 30-seconds to convey the message.

Possible ideas:

____

____

____

____

____

____

Select the best idea and create a 30-second script. Remember that a television script includes both audio and visual directions and effects.

Script		
Character or Event	**Dialog or Directions**	**Production Notes**
Example - - -		
Musical score	*Soft mellow music*	*Fade out to dialog*
Announcer	*Bob and Carol were walking home from school on a windy day.*	
Sound effects	*Blowing wind*	*Background growing in intensity*
	Start your script here and complete on next page.	

Name ___________________________________

Script–Page 2		
Character or Event	**Dialog or Directions**	**Production Notes**

Script–Page 3		
Character or Event	**Dialog or Directions**	**Production Notes**

Name ______________________ Date ____________ Class ______________

ENGINEERING & TECHNOLOGY ACTIVITY

Taking Your Technology Knowledge Home: Planning a Trip

Many people use their own vehicles to travel from one place to another. Discuss the possibilities of taking a driving trip with your parents or guardians that will fulfill a specific purpose, but also allow room for sightseeing and other brief stops. Prepare a possible plan for this trip by completing the following steps.

Step 1. Identify the origin (home), destination(s), and time that can be allowed for the trip.

Origin: ______________________

Destination: ______________________

Length of trip: ______ days

Step 2. Identify the main purpose of the trip using a statement such as "Visit my grandmother" or "See the Grand Canyon."

Purpose: ______________________

____

____

Step 3. List any special sights, events, locations, or attractions that could be included on the trip, such as sporting events, theme parks, historic sites, national parks, or other special places. Indicate the amount of time you would like to spend at each one.

Stop: ______________________

Time allotted: ______________________

Stop: ______________________

Time allotted: ______________________

Stop: ______________________

Time allotted: ______________________

Stop: ______________________________

Time allotted: ______________________________

Stop: ______________________________

Time allotted: ______________________________

Step 4. Use road maps or the Internet to determine various routes for the trip. Consider all the places you want to visit and stops you plan to make, as well as the total amount of time you have for the trip. Determine the best directions and itinerary for your trip and record it in the space provided. An example entry for the first day might be, "Day 1, 8:00 a.m. Leave home and drive 30 miles on Route 34 to Newtown. Arrive at 9:00 a.m." Plan each day's events on separate lines.

Name ___________________________________

Step 5. Identify potential expenses for each part of the trip. Prepare a cost estimate for the trip by completing the following chart. Present this chart and your proposed itinerary to your parents or guardians.

Starting point: ______________________ Destination: ______________________

Starting date: ____________ Ending date: ____________ Round trip miles: ________

Estimated Costs:

Gasoline: $______

Food: $______

Lodging: $______

Event and sight-seeing fees: $______

Total: $ ______

Name ____________________ Date __________ Class ______________

ENGINEERING & TECHNOLOGY ACTIVITY

Taking Your Technology Knowledge Home: Investigating Lighting Costs and Savings

Energy costs are on the rise. Many homeowners are saving money by replacing standard (incandescent) light bulbs with light-emitting diode (LED) light bulbs. In this activity, you will survey lighting in at least three rooms of your home and calculate the savings that can be obtained by replacing incandescent bulbs with energy-efficient LED bulbs.

Evaluate the Existing Lighting

Step 1. Locate, identify, and list all the lighting in at least three rooms. Enter the information in the following chart.

Present Lighting						
Room	Type of Lamp	Wattage	Hours per Month	Kilowatt-Hours (kWh) per month	Cost per kWh	Cost per Month

Step 2. Find the wattage rating for each fixture or lamp. You can usually find it by looking at the fixture, lamp, or bulb. Enter the value in the "Wattage" column.

Step 3. Estimate how many hours each lamp is used each day. Multiply the hours used per day by 30 to obtain an estimate of hours used per month. Enter this number in the "Hours per Month" column.

Step 4. Determine the kWh each light uses by multiplying the wattage by the hours used per month. The result is in watt-hours of power used in one month. To find the kilowatt-hours per month, divide the answer by 1000. Enter the value in the "Kilowatt-Hours (kWh) per Month" column.

Step 5. Find the cost of a kWh of electricity in your city. This information can often be found on your utility bill or the utility company's website. Enter the amount in the "Cost per kWh" column.

Step 6. Multiply the kWh used by the cost per kWh. Enter the amount in the "Cost per Month" column.

Step 7. Total all the entries in the "Cost per Month" column to obtain the current, estimated lighting costs for the rooms surveyed.

$ ____________________

Proposed Cost Savings

Now, calculate the possible energy and cost savings of replacing all the incandescent light bulbs with energy-efficient LED bulbs.

Step 1. In the following chart, list the same rooms and types of lamps or fixtures you included in the first chart.

LED Lighting

Room	Type of Lamp	Wattage	Hours per Month	kWh per Month	Cost per kWh	Cost per Month

Step 2. Select a proper replacement LED bulb for each incandescent bulb. Enter the wattage value in the chart.

The recommended replacement bulbs are as follows:

- 40-watt standard bulb = 7-watt LED bulb
- 60-watt standard bulb = 10-watt LED bulb
- 75-watt standard bulb = 12-watt LED bulb
- 100-watt standard bulb = 18-watt LED bulb
- 150-watt standard bulb = 27-watt LED bulb

Step 3. Calculate the monthly kWh each lamp uses by multiplying the wattage by the hours the light is used each month. Divide the answer by 1000 to convert the watt-hours into kWh. Enter the amount in the "kWh per Month" column on your chart.

Step 4. Enter the cost of a kWh of electricity in the "Cost per kWh" column on your chart.

Step 5. Multiply the kWh used by the cost per kWh. Enter the amount in the "Cost per Month" column on your chart.

Step 6. Total the "Cost per Month" column to obtain your estimated cost of lighting using energy-efficient bulbs.

$ ____________________

Calculating Savings

Step 1. Subtract the cost of operating the LED lighting from the cost of operating the standard lighting. This is the savings per month.

$ ____________________

Step 2. Determine the cost of replacing the bulbs by finding prices at a local store, on the Internet, or from prices provided by your instructor.

$ ____________________

Step 3. Divide the cost of the bulbs by the monthly savings to determine the number of months it will take to pay for the cost of replacing the bulbs. LED bulbs last about fifty times longer than incandescent bulbs. This longer life helps defray the cost of replacing the bulbs.

____________________ months to recover replacement cost

Name ______________________ Date __________ Class ______________

ENGINEERING & TECHNOLOGY ACTIVITY

Taking Your Technology Knowledge Home: Designing a Tool

Introduction: Tools are important to all people. Tools are involved in almost every task we do. For example, we use writing tools (pencils and pens), measuring tools (rulers and tape measures), kitchen tools (knives, mixing spoons, and measuring cups), and garden tools (trowels and hoes), to name a few. Think of all the tools you use in one day.

Challenge: Design a tool that can be used around your home or for doing yard work.

Step 1. Identify a task for which you can design a tool.

____

____

Step 2. Look through magazines, textbooks, and the Internet to find tools that perform similar tasks. List the important features of the tools you find.

____

____

____

____

____

____

Step 3. List the features your tool will have.

____

____

____

____

Step 4. On the following grid, sketch four possible designs for the tool you plan to make.

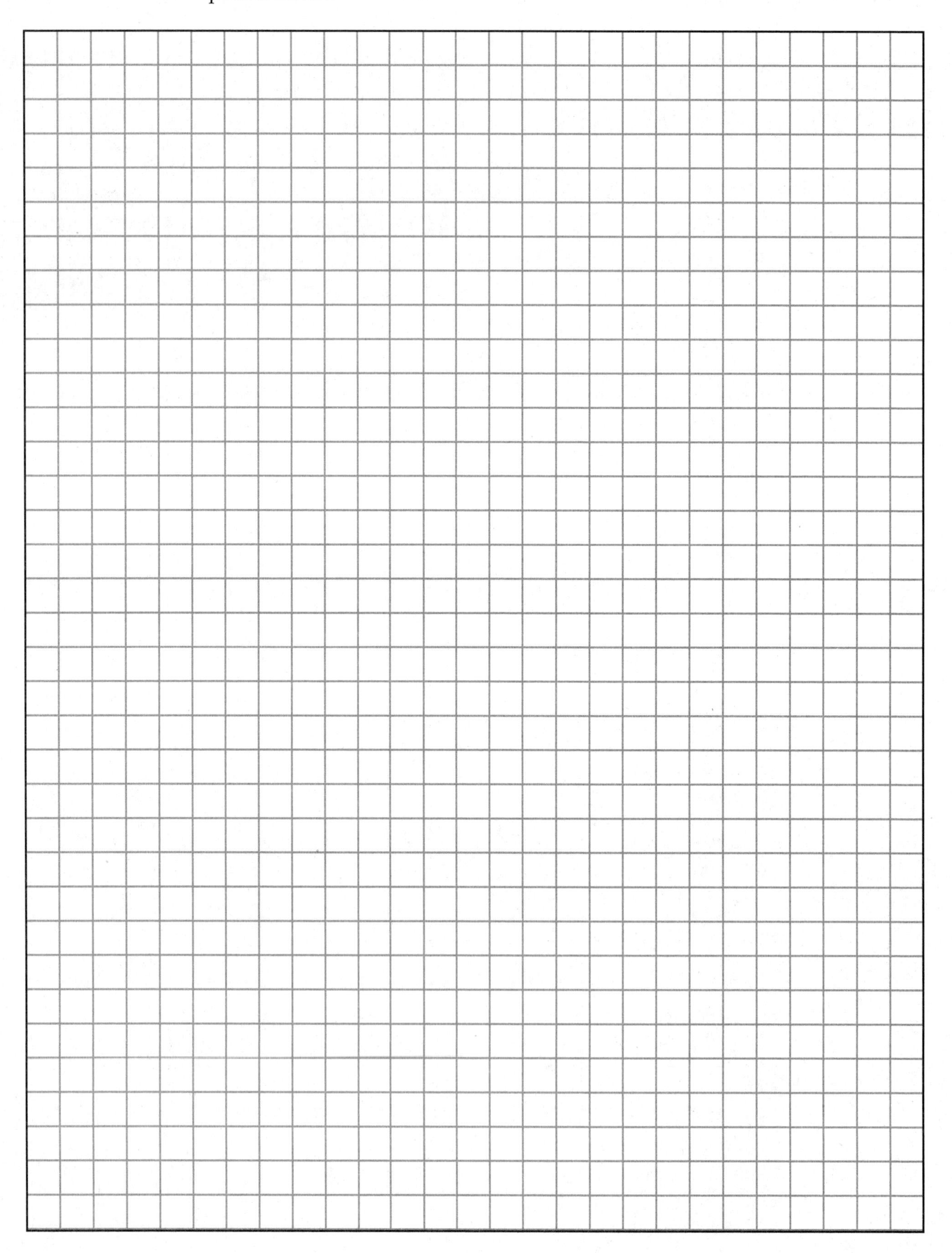

Step 5. Select your best design. On the following grid, prepare a better (refined) sketch of the tool.

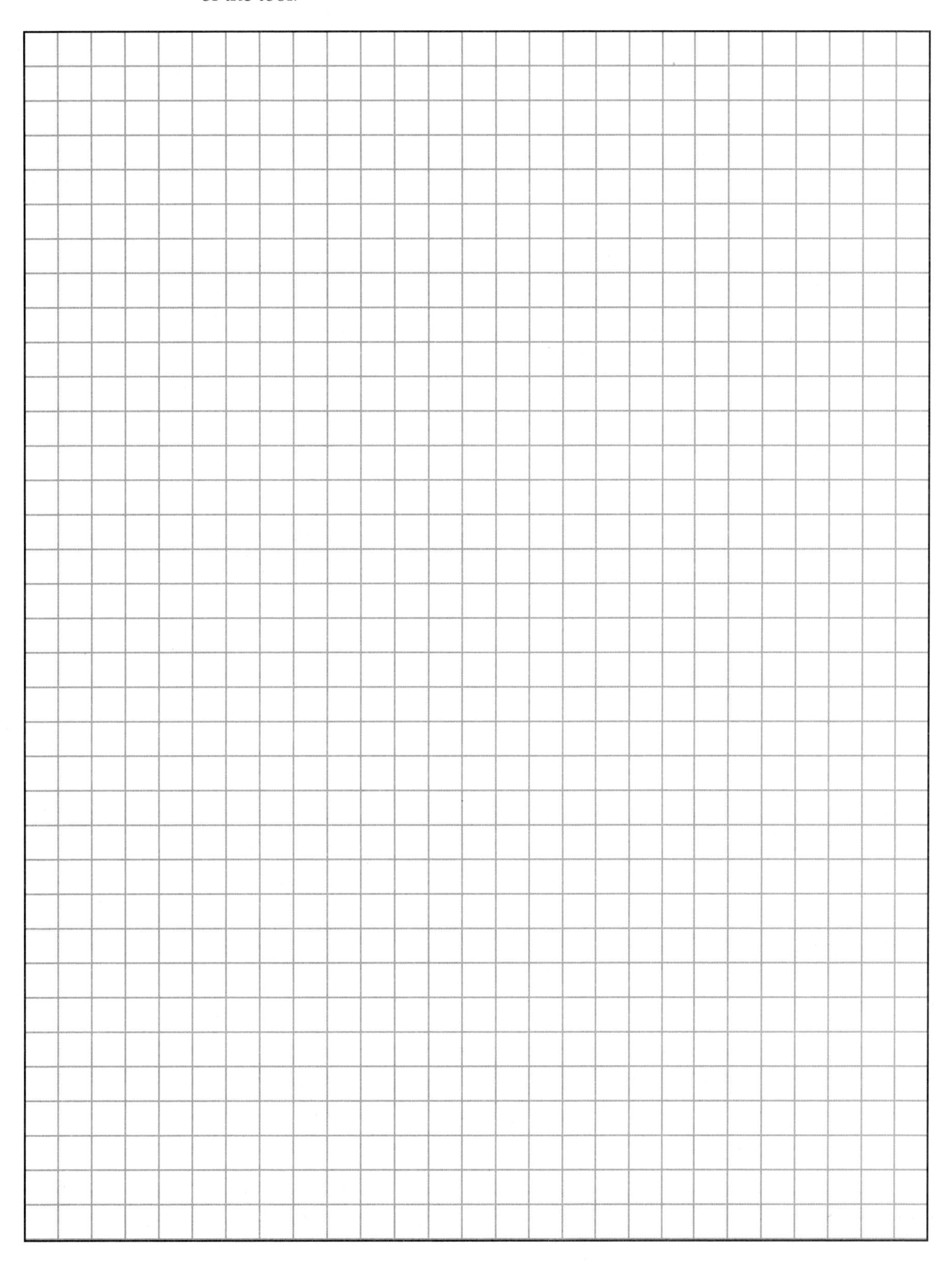

Step 6. Build a sample of the tool.

Step 7. Have someone test the tool and comment on its design. Record the comments you received in the space provided.

____

____

____

____

____

____

Step 8. On the following grid, sketch an improved design for your tool that incorporates suggestions from the tester.

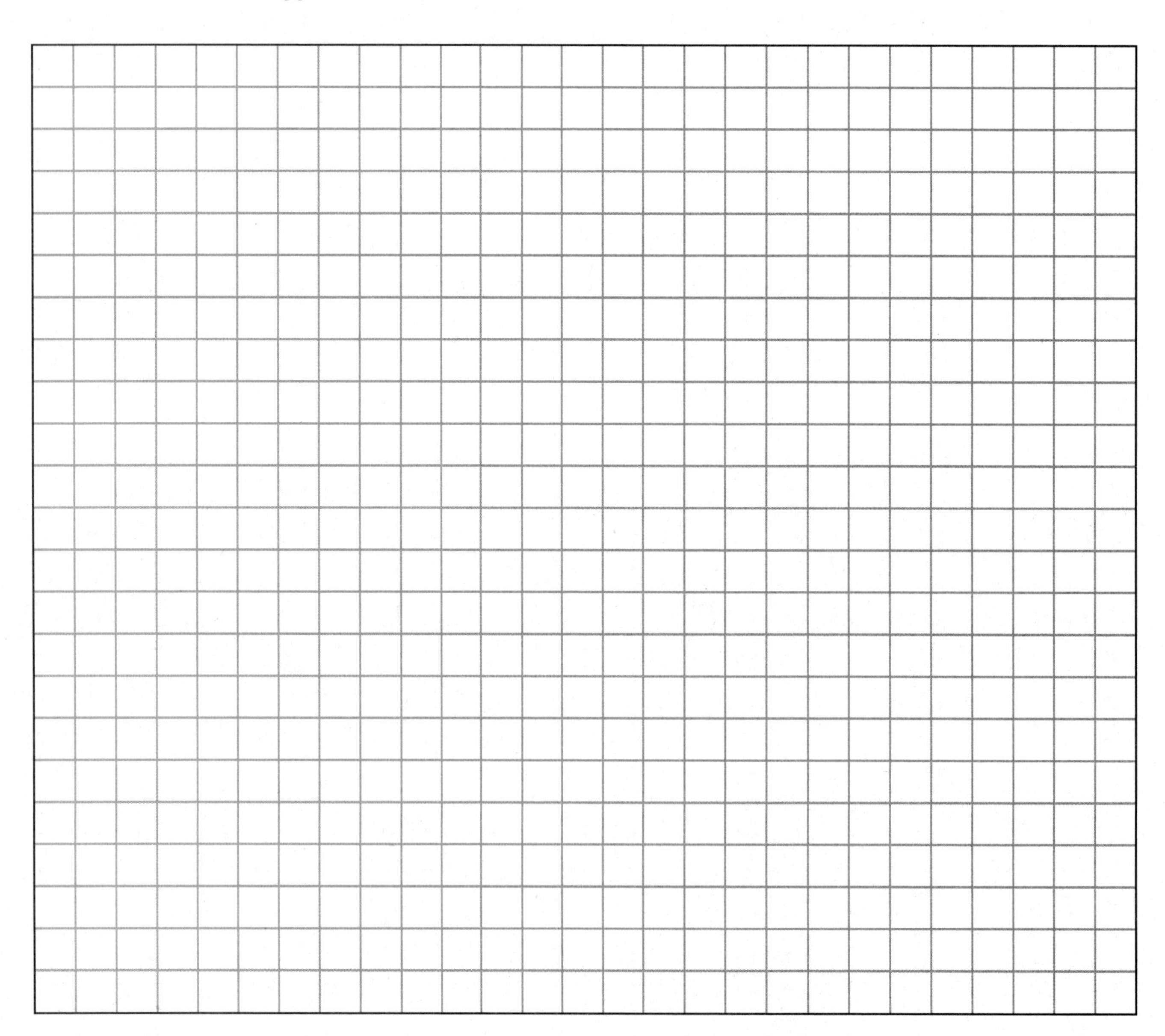

Name ______________________ Date __________ Class ______________

ENGINEERING & TECHNOLOGY ACTIVITY

Taking Your Technology Knowledge Home: Designing a Manufactured Product

Challenge: Design a product that you and your friends can manufacture for sale or give to a charity.

Step 1. Identify the use for the product (circle one).

- Sale
- Charity

Step 2. Identify the type of product it will be (circle one).

- Toy
- Game
- Household item
- Office item
- Yard item
- Other: ____________

Step 3. Specifically define the function of the product. For example, a device to hold headphones or a game to entertain children on a trip.

____

____

____

____

Step 4. List the features the product will have.

____

____

____

____

Step 5. List the limitations you must consider, such as cost, materials, and available tools.

____

____

____

____

Step 6. On the following grid, sketch four possible designs for the product you plan to make.

Step 7. Select your best design and prepare a refined sketch of the product on the following grid.

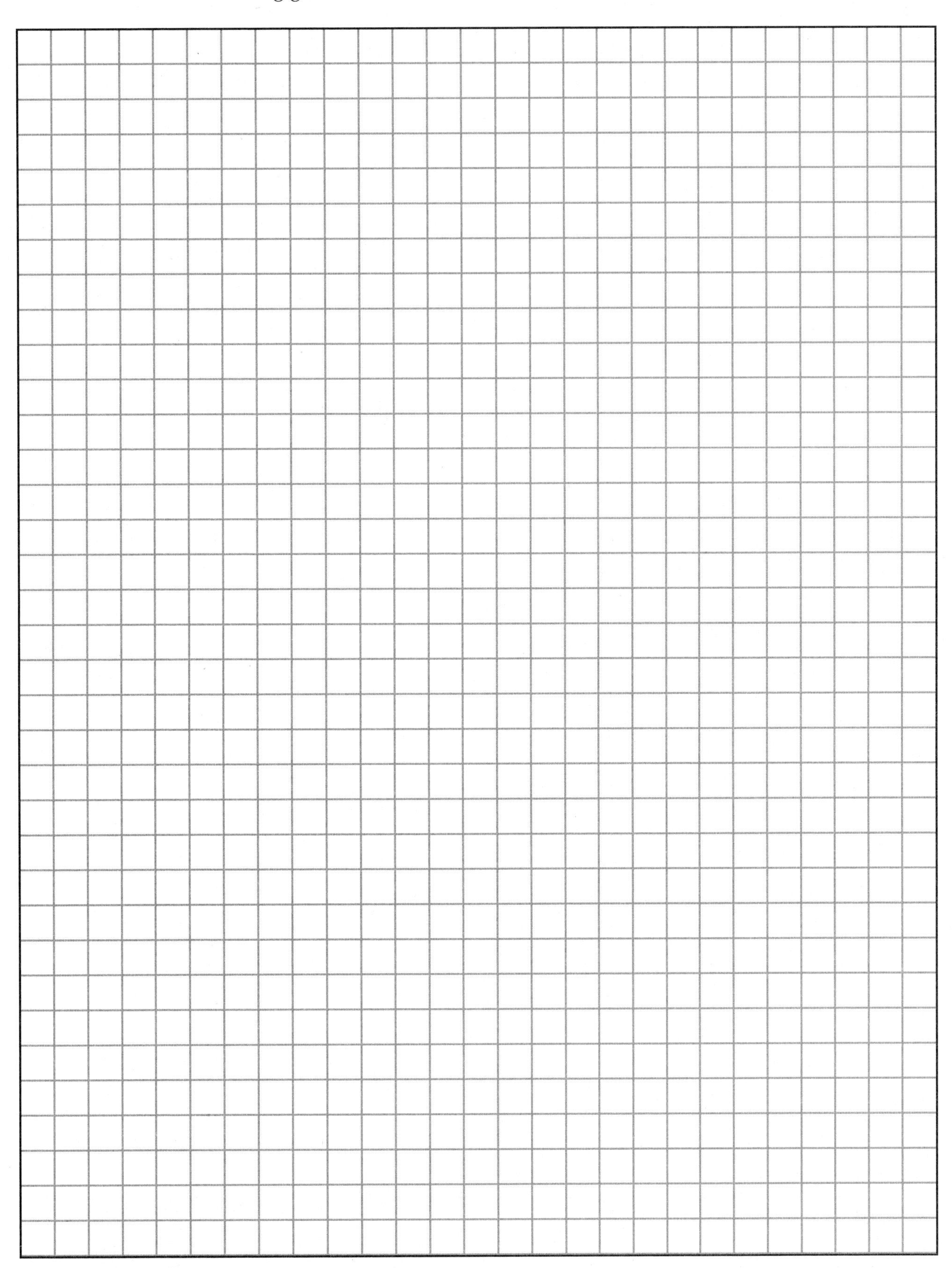

Step 8. Build a sample of the product.

Step 9. List the procedure you would follow to make 10 of the product.

____
____
____
____
____
____
____
____
____
____
____
____

Step 10. On the following grid, design a package for the product.

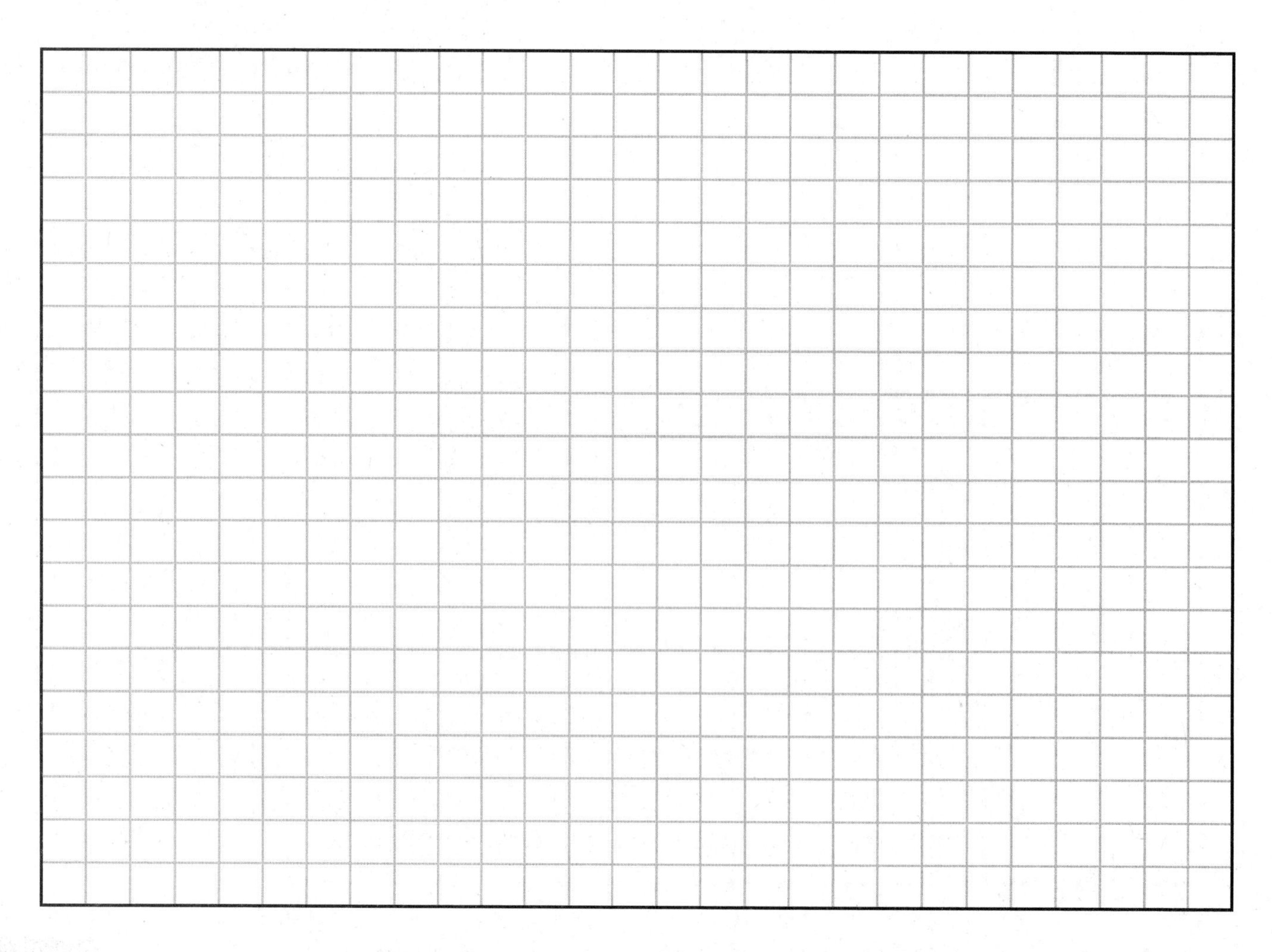

Name ______________________ Date __________ Class ______________

ENGINEERING & TECHNOLOGY ACTIVITY

Taking Your Technology Knowledge Home: Practicing Primary and Secondary Food Processing

People practice both primary and secondary food processing on a regular basis. Some people grind their own flour, make their own sausage, or can fresh vegetables—all are examples of primary processing. Later, they use the results of their primary processing efforts to produce meals, which is a secondary process.

In this activity, you will practice primary and secondary processing activities at home. You will make peanut butter using primary processing techniques, and put the peanut butter on a biscuit that you bake using secondary processing techniques.

Primary Food Processing

Primary food processing is the process of changing an agricultural product into a food ingredient. Peanuts (agricultural product) are used to make peanut butter (food ingredient). Peanut butter is used to make a number of different products, including snack cracker sandwiches, cookies, and ice cream. To make peanut butter using primary food-processing techniques, complete the following steps.

Step 1. Determine the type of peanut butter you will make (circle one).

- Creamy
- Crunchy

Step 2. Collect the following ingredients:

- Roasted peanuts in shells
- Corn or vegetable oil
- Sugar

Step 3. Gather the following equipment:

- Tablespoon
- Teaspoon
- Measuring cup
- Food processor or grinder (for crunchy-style peanut butter)
- Food blender (for creamy-style peanut butter)
- Clean, empty jar

Step 4. Place a cup of peanuts in a food processor or grinder.

Step 5. Add 1/2 teaspoon of peanut or vegetable oil.

Step 6. Grind the nuts for two to three minutes. Check the consistency.

Step 7. Add additional oil or grind longer to obtain the desired consistency (creamy or chunky).

Step 8. Add sugar slowly, until the desired taste is obtained.

Note: Peanut butter contains no preservatives and must be stored in the refrigerator.

Have someone taste your peanut butter and give you feedback. Record the feedback in the space provided. Make adjustments to the peanut butter based on the feedback and your own taste test.

____

____

____

____

Secondary Food Processing

Secondary food processing is the process of changing food ingredients into edible products. Biscuits are bread products produced from various food ingredients. To make a batch of biscuits using secondary food-processing techniques, complete the following steps.

Step 1. Collect the following ingredients:

- 2 cups flour
- 1/2 teaspoon cream of tartar
- 1 tablespoon baking powder
- 1/4 teaspoon salt
- 2 teaspoons sugar
- 1/2 cup butter, shortening, or margarine
- 3/4 cup buttermilk

Step 2. Gather the following equipment:

- Tablespoon
- Teaspoon
- Measuring cup
- Sifter or mesh strainer
- Mixing bowl
- Mixing spoon
- Baking or cookie sheet
- Biscuit cutter or drinking glass

Step 3. Set the oven to 450°F.

Step 4. Sift the flour, baking powder, sugar, cream of tartar, and salt together in a mixing bowl.

Step 5. Add the butter, shortening, or margarine in small chunks and combine until the mixture resembles coarse crumbs.

Step 6. Make a well in the center of the mixture.

Step 7. Pour the buttermilk into the well all at once.

Step 8. Stir just until the dough clings together.

Step 9. Place the dough on a lightly floured surface and flatten to about 1/2″ thick.

Step 10. Cut disks of dough using a biscuit cutter or drinking glass. Dip the cutter in flour between cuts.

Step 11. Transfer the biscuits to a baking sheet.

Step 12. Place the baking sheet in the preheated oven. Bake for 10–12 minutes or until biscuits are golden.

Have someone taste your biscuits and give you feedback. Record the feedback in the space provided.

____

____

____

____

Name ______________________ Date __________ Class ______________

ENGINEERING & TECHNOLOGY ACTIVITY

Taking Your Technology Knowledge Home: Designing a Code

Introduction: Codes are used for many things. Manufacturers put codes on products indicating when and where products were made. Most products have bar codes that are read by optical scanners to check inventory and to read the price of products at checkout counters. QR codes are used to direct users to websites, information, or other resources. Codes are also used to send messages. Morse code is an example of a message code. This code uses a series of dots and dashes to represent the letters of the alphabet. The codes are sent as electrical impulses over telegraph wires, as audible sounds over shortwave radio, or as flashes of light in ship-to-ship communication. Language is another code for sending a message. Sounds we make with our mouths and lips represent ideas and things. For example, different sounds represent the object "desk" in different languages.

Code Talkers

One unique use of language codes appeared during World War II. The Navajo language was used as a code throughout the campaign in the Pacific. Code talkers, U.S. servicemen of Navajo descent, transmitted voice messages in their native language. The enemy forces never broke this code.

The success of the language as a code rested on the fact that the Navajo language is a very complex, unwritten language. The sounds and syntax of the language make it difficult to interpret without extensive exposure and training. In addition, the Navajo language is not a written language, so it has no associated alphabet or symbols.

The Navajo code talkers developed a code using their words to express military terms. A message was a string of seemingly unrelated Navajo words. To translate the message, the code talker translated the Navajo words into English equivalents. (For example, the Navajo word *wol-la-chee* means "ant.") The code talker then took the first letter of each English word to form the decoded word. A message might contain the Navajo words, *tsah*, *wol-la-chee*, *ah-keh-diglini*, and *tsahah-dzoh*. The words are translated into "needle," "ant," "victor," and "yucca." Taking the first letter of each English equivalent, the word *navy* appears.

To make the code easier to send, however, some common terms were assigned a specific Navajo word. For example, the Navajo word *dah-he-tih-hi* means "hummingbird" and was used to mean "fighter plane." While enemy forces were able to break many codes used in the war, none were able to break the Navajo code-talker code used by the Marines.

Code Systems

Code systems are used in a number of different situations. If you have ever listened to a police scanner or watched a police-type TV show, you have heard code being used. The code "10-4" means "okay," while "10-19" means "return to the station." There is an entire system of "10" codes used nationwide by police and fire dispatchers to communicate with personnel in the field. Another example is train-whistle signal code. Two short blasts of the whistle mean the engineer plans to start a standing train. Three short blasts mean the engineer plans to back up a standing train. There are many other signals engineers use to indicate their planned actions.

The following drawing illustrates three other examples of simple codes. These and all other codes have two things in common:

- There is a way to shorten or hide the meaning of the message (signs or symbols).
- There is a way to send the message (technology).

The technology involved might use visual devices, such as flags, flashes of light, or printed symbols, or electrical and electronic equipment, such as the telegraph or the radio.

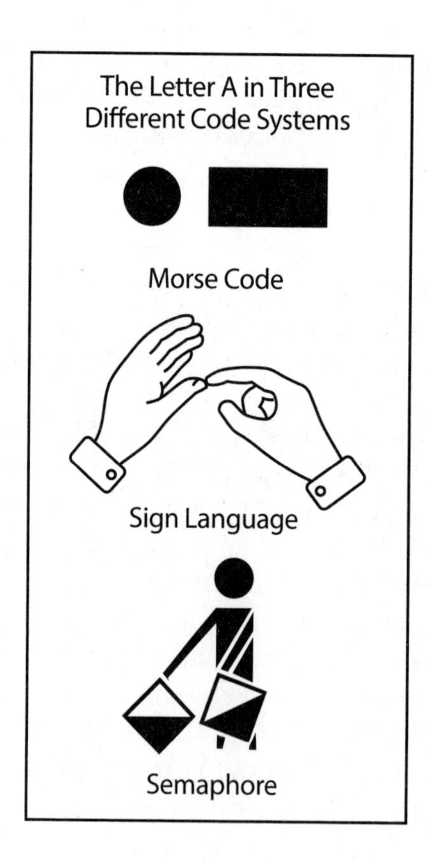

Name ________________________

Challenge: Develop a set of symbols, or codes, that can be used to communicate a message using technology (equipment or devices) to your best friend without other people understanding the message.

Step 1. Decide how the message will be sent, such as through visual images, flashes of light, electronic signals, or audio sounds.

____

____

____

Step 2. Develop a code for each letter and number.

A. ☐	B. ☐	C. ☐	D. ☐	E. ☐
F. ☐	G. ☐	H. ☐	I. ☐	J. ☐
K. ☐	L. ☐	M. ☐	N. ☐	O. ☐
P. ☐	Q. ☐	R. ☐	S. ☐	T. ☐
U. ☐	V. ☐	W. ☐	X. ☐	Y. ☐
Z. ☐				
1. ☐	2. ☐	3. ☐	4. ☐	5. ☐
6. ☐	7. ☐	8. ☐	9. ☐	0. ☐

Step 3. On the following grid, sketch the technical system you will use for your communication system (such as an electronic circuit or a flag construction).

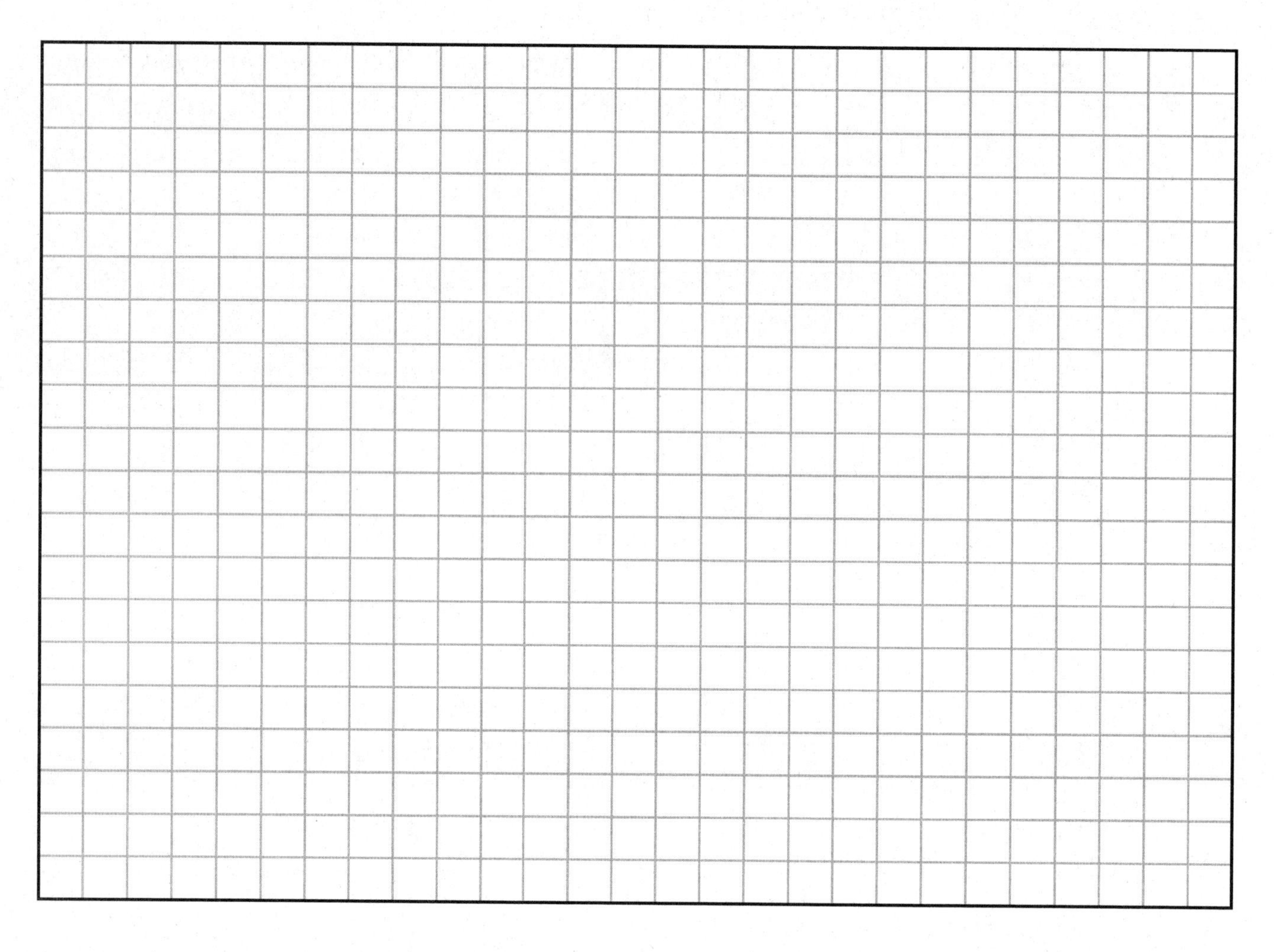

Step 4. Code the phrase "Can you read this" in the following boxes.

C	A	N	Y	O

U	R	E	A	D

T	H	I	S

Name ______________________ Date __________ Class ______________

ENGINEERING & TECHNOLOGY ACTIVITY

Taking Your Technology Knowledge Home: Promoting a Neighborhood Event

The Internet and online video services have significantly impacted the way people receive, view, and process news and information. Many people are more familiar with popular online videos than current events or local activities.

You have been asked to create a 30-second video to promote a neighborhood event. To do this, complete the following steps.

Step 1. Identify an event, the sponsor, and contact person.

Event: ______________________________

Sponsor: ______________________________

Contact person: ______________________________

Step 2. Gather information about the event.

Date: ______________________________

Time: ______________________________

Location: ______________________________

Cost to attend: ______________________________

Restrictions: ______________________________

Purpose: ______________________________

Other information to be conveyed: ______________________________

____

____

Step 3. Develop a theme for the video.

____

Step 4. Develop a storyline and important message for the video.

____

____

____

Step 5. Develop a possible script for the video. Sketch each scene using the following grid.

Step 6. Present the sketches to the sponsor and ask for feedback and suggestions. Record the sponsor's comments in the space provided.

____

____

____

____

Step 7. Produce the 30-second video.

Name ______________________ Date __________ Class ______________

ENGINEERING & TECHNOLOGY ACTIVITY

Taking Your Technology Knowledge Home: Developing an Organization Management System

Groups, businesses, and individual people are constantly organizing events, items, and activities. Each of these can be better managed if they are organized efficiently and effectively. You can help others by developing a management system for organizing important events, items, or activities.

To do this, complete the following steps.

Step 1. Identify an event (e.g., bake sale), item (e.g., movie collection), or activity (e.g., sports tournament) for which you can develop a management plan.

Event/item/activity: ______________________

Sponsor/owner/participant: ______________________

Step 2. Meet with the sponsor, owner, or participant to discuss the need for a management structure.

What needs to be managed? ______________________

____

How many items/events/activities need to be managed? ______________________

____

What possible categories could be used to manage the items/events/activities?

____

How have these items/events/activities been managed in the past?

____

____

Step 3. On the following grid, develop a management organization system for the event/item/activity you have identified and discussed.

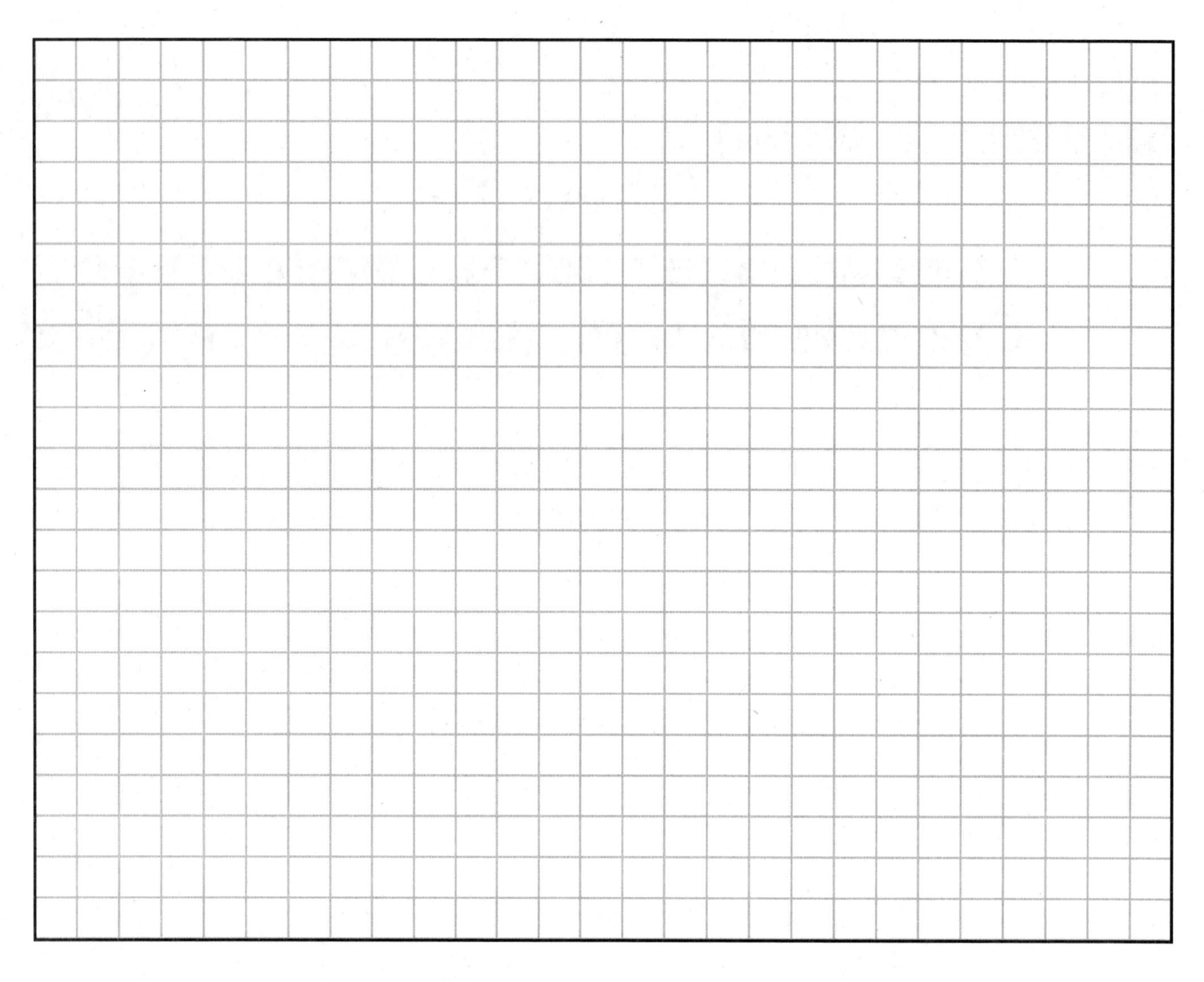

Step 4. Present management organization system to the event sponsor, owner, or participant. Record their comments and suggestions in the space provided.

____

____

____

____

Step 5. Using the feedback, revise your management system and sketch the new system.

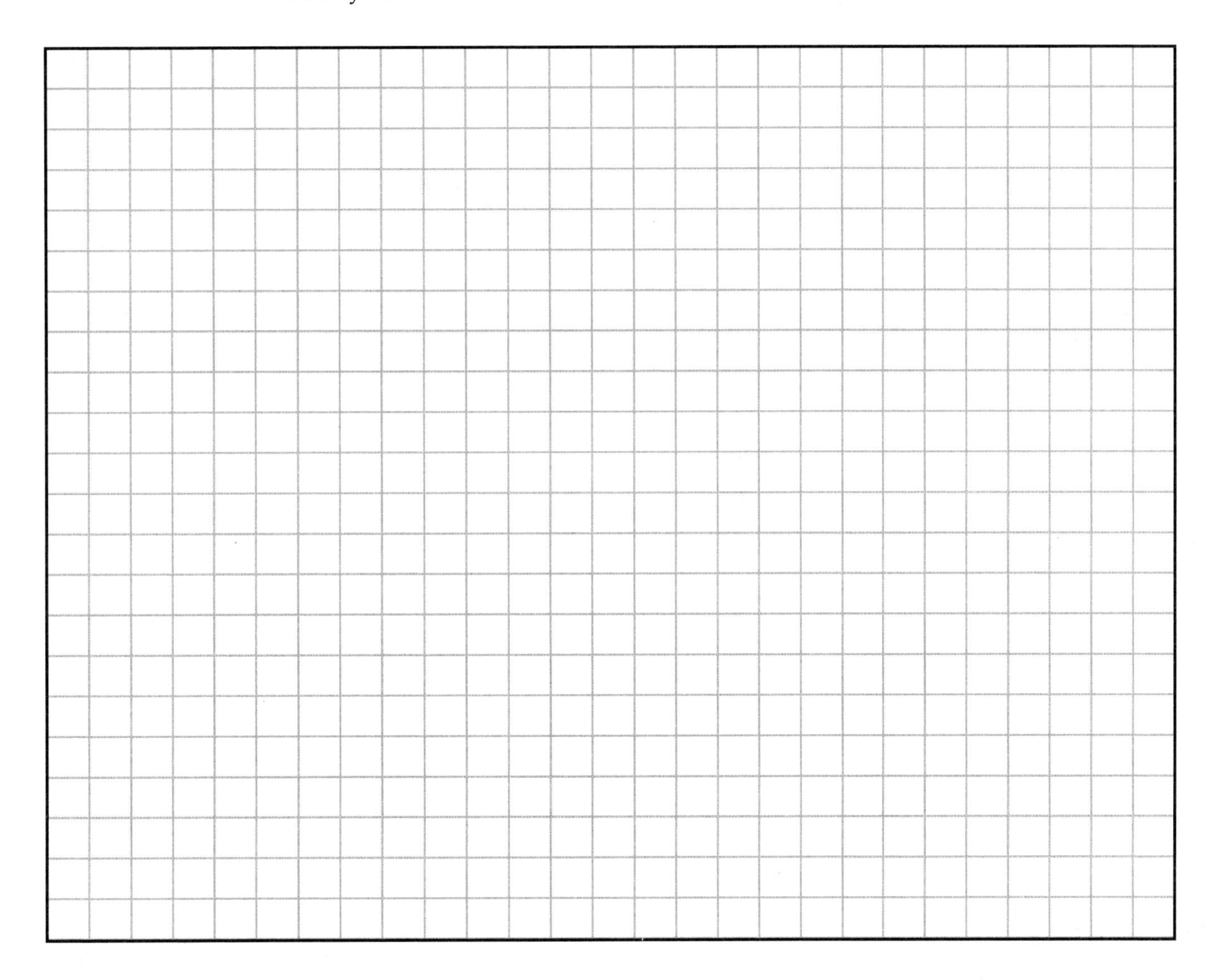